CIRIA C634 London 2005

AF744477

Management of accelerated low water corrosion in steel maritime structures

Prepared under contract to CIRIA by Mott MacDonald in partnership with BAC Corrosion Control Ltd, Nuttall John Martin and the Corrosion and Protection Centre at the University of Manchester

J E Breakell BSc, MSc, PhD, CEng, CSci, MIChemE

M. Siegwart BEng, Dipl-Ing(FH), PhD, Eur Ing, CEng, MIMMM, MICorr

K. Foster BTech, CEng, MICE

D Marshall BSc, PhD, MICorr

M Hodgson BSc, CEng, MICE, MIStruct.E

R Cottis BA (Hons), MA, PhD

S Lyon MA, PhD, DSc, CEng, FIMMM, FICorr

THE UNIVERSITY OF BOLTON
THE LIBRARY EAGLE CAMPUS
WITHDRAWN
Accession No. 389953
Classification No. 624.18341 BRE

CIRIA sharing knowledge ▪ building best practice

Classic House, 174–180 Old Street, London EC1V 9BP
TELEPHONE 020 7549 3300 FAX 020 7253 0523
EMAIL enquiries@ciria.org
WEBSITE www.ciria.org

THE UNIVERSITY OF BOLTON
389953

10175262

Summary

ALWC is an aggressive and localised form of low water corrosion which can occur in tidal and brackish waters on steel maritime structures. Corrosion rates in regions affected by ALWC can be 1 mm/side/year or more and design corrosion allowances will, therefore, be exceeded quickly. This process will rapidly compromise the integrity of affected structures and will lead to significant costs for repair or replacement, in addition to having serious implications for the safe operation of the structure.

This book provides a comprehensive guide to the phenomenon of ALWC and its management, condition appraisal, repair, protection and monitoring. It is aimed at infrastructure owners, operators and their advisors, specialist engineers and asset and maintenance managers in ports, harbours and other marine locations. An executive summary gives an overview of the book and answers key questions. More in-depth information can be gained from reading the whole book, particular chapters of interest or the detailed information in the appendices.

The book is based on a detailed review of published literature and infrastructure owner's procedures, consultation with experts and practitioners from a cross-section of fields of expertise and case studies demonstrating good practice.

Management of accelerated low water corrosion in steel maritime structures

Breakell J; Siegwart M; Foster K; Marshall D; Hodgson M; Cottis R; Lyon S

CIRIA C634 © CIRIA 2005 ISBN 0-86017-634-7 RP693

British Library Cataloguing in Publication Data

A catalogue record is available for this book from the British Library.

<table>
<tr><td colspan="3">Keywords
Asset management, coastal and marine, water infrastructure</td></tr>
<tr><td>Reader interest
Infrastructure owners and operators, and their advisors, specialist engineers (experts in marine activities), asset and maintenance managers</td><td>Classification
AVAILABILITY
CONTENT
STATUS
USER</td><td>
Unrestricted
Enabling document
Committee-guided
Asset managers, maintenance, maritime, structural and civil engineers</td></tr>
</table>

Published by CIRIA, Classic House, 174–180 Old Street, London EC1V 9BP, UK.

All rights reserved. No part of this publication may be reproduced or transmitted in any form or by any means, including photocopying and recording, without the written permission of the copyright-holder, application for which should be addressed to the publisher. Such written permission must also be obtained before any part of this publication is stored in a retrieval system of any nature.

This publication is designed to provide accurate and authoritative information in regard to the subject matter covered. It is sold and/or distributed with the understanding that neither the author(s) nor the publisher is thereby engaged in rendering a specific legal or any other professional service. While every effort has been made to ensure the accuracy and completeness of the publication, no warranty or fitness is provided or implied, and the author(s) and publisher shall have neither liability nor responsibility to any person or entity with respect to any loss or damage arising from its use.

Acknowledgements

This CIRIA book *Management of accelerated low water corrosion in steel maritime structures* was produced as a result of Research Project RP693 which was carried out under contract to CIRIA by Mott MacDonald Ltd in partnership with BAC Corrosion Control Ltd, Nuttall John Martin and The Corrosion & Protection Centre at the University of Manchester.

The book sets out good practice in the management, condition assessment, repair, protection and monitoring of steel maritime structures affected by ALWC.

Following CIRIA's usual practice, the research project was guided by a Steering Group which comprised:

Mr I Wesley (chairman)	W S Atkins Ltd
Dr I Beech	University of Portsmouth
P Beckett	Associated British Ports
J Christie	Aberdeen Harbour Board
E Marsh	Corus Ltd.
R James	iicorr Ltd
D Kent	Edmund Nuttall Ltd
P Lacey	LA Consult
A Martin	Arch Henderson
B McCann	British Ports Association/Fraserburgh Harbour
A McKay	Port of London Authority
M Roe	W S Atkins (DTI representative)
I Schofield	Associated British Ports
M Shaw	Kirk McClure Morton
T Shelley	Shelley Consulting Services
B Wyatt	Corrosion Control

Corresponding members comprised:

Prof. R Akid	Sheffield Hallam University
C Boysons	Halcrow Group Ltd
M Briers/ D Kavanagh	Environment Agency
D Deacon	Institute of Corrosion
G Gedge	Arup Materials Consulting
P Martin	Peter Fraenkel Maritime
J Perry	Mott MacDonald Ltd
J Service	Jacobs Babtie
Dr J D Scantlebury	University of Manchester
P Shone	Chevron Texaco Ltd
J Simm	HR Wallingford
P Wright	Royal Haskoning

CIRIA's Research Manager was Ms S Reid. The project was initiated by Ms E Holliday.

Other organisations contributed to the book by reviewing drafts and specific text. They include the British Ports Association Engineering Dredging Group, Arup worldwide via their intranet, and the ICE Maritime Board.

Contributions do not imply that individual funders, contributors or Steering Group members necessarily endorse all views expressed in this publication.

This project was funded by the Department of Trade and Industry, Environment Agency, Nuttall John Martin, Van Ord ACZ, Arch Henderson, iicorr Ltd, Port of London Authority, Peter Fraenkel Maritime Ltd, Chevron Texaco and by CIRIA's Core Programme Sponsors.

Unattributed photographs have kindly been provided by the authors and Steering Group Member organisations.

The Library

The University of Bolton
Deane Campus
Deane Road
Bolton, BL3 5AB

This book is available for loan to registered borrowers only

WITHDRAWN

Classification No 624.183 BRE 41

Accession No 389953

Health and Safety

Construction activities, particularly those on construction sites, have significant health and safety implications. These can be the result of the activities themselves or can arise from the nature of the materials and the chemicals used in construction. This book gives some reference to relevant health and safety issues. However, other published guidance on specific health and safety issues in construction should be consulted as necessary to ensure up-to-date legislation is applied and appreciated, especially the requirements of national legislation and those of infrastructure owners.

Dr. Jim Breakell, BSc, MSc, PhD, CEng, CSci, MIChemE

Dr Breakell is a charted chemical engineer at Mott MacDonald Special Services and is a recognised expert in corrosion and materials engineering. He is a leading figure in pioneering developments in the understanding and remediation of ALWC on steel structures and is the principal author of this publication.

Dr. Michael Siegwart, BEng, Dipl.-Ing(FH), PhD, Eur Ing, CEng, MIMMM, MICorr

Dr Siegwart is a chartered materials engineer at Mott MacDonald Special Services. He has published several papers on electrochemical repair in international peer reviewed journals and has practical experience in the electrochemical rehabilitation of marine structures.

Mr. Keith Foster, BTech., CEng, MICE

Mr Foster is a chartered civil engineer at Mott MacDonald Maritime. He has more than 25 years' experience of project management, survey, design, specification and construction of maritime structures, with particular experience of remedial and upgrading works. He is also a planning supervisor under the Construction (Design and Management) Regulations.

Dr. Dan Marshall BSc, PhD, MICorr

Dr Marshall is a senior project engineer at BAC Corrosion Control Limited. He is experienced in the design and survey of cathodic protection systems, both galvanic and impressed current and has installed numerous cathodic protection systems in marine facilities.

Mr. Mike Hodgson, BSc, CEng, MICE, MIStruct.E

Mr Hodgson is technical manager at Nuttall John Martin. He is an acknowledged expert in the field of repair to underwater structures, particularly those suffering from ALWC and in the development of moveable cofferdams (limpet dams) to gain dry working access below water level.

Dr. Bob Cottis, BA(Hons), MA, PhD

Dr Cottis is a Reader in Corrosion Science and Engineering within the Corrosion and Protection Centre at the University of Manchester. His major areas of study have included corrosion fatigue, hydrogen embrittlement, pitting corrosion (especially statistical aspects), the theoretical aspects of electrochemical noise measurement and analysis and the application of neural network and other computing and information technology methods to the modelling and control of corrosion.

Dr. Stuart Lyon, MA, PhD, DSc, FIMMM, FICorr, CEng

Dr Lyon is a Reader in Corrosion Science and Engineering at the Corrosion and Protection Centre at the University of Manchester. His research interests include: accelerated testing and analysis of materials performance, advanced paints and coatings for materials protection, engineering failure analysis and advanced electrochemical methods.

Executive summary

Who is the book for? The target audience is:

- port infrastructure owners and operators and their advisors
- specialist engineers – experts in marine activities
- asset and maintenance managers.

Owners and operators of other structures such as coastal defences will also find the book valuable.

What is it about? Accelerated Low Water Corrosion (ALWC) is a particularly aggressive and localised form of low water corrosion which can occur in tidal and brackish waters on steel maritime structures. This book aims to provide a comprehensive guide to the phenomenon of ALWC and its management in ports and harbours and other marine locations.

Why should I read this book? Corrosion rates in regions affected by ALWC can be 1 mm/side/year or more and design corrosion allowances will, therefore, be exceeded quickly. ALWC could lead to a loss of 33 to 66 per cent of the asset value. This process will rapidly compromise the integrity of affected structures and will lead to significant costs for repair or replacement, in addition to having serious implications for the safe operation of the structure.

How is this book structured? You do not have to read this guide in any particular order. It is structured so that you can obtain as little or as much information as you may need on ALWC and its management. Although this approach is designed to help the reader, it inevitably results in some repetition.

A first level of understanding can be gained by reading this summary in combination with those found at the end of each of the main chapters and Chapter 10 "Recommendations for good practice". Greater understanding can be obtained from reading the whole book or those chapters of interest which are augmented by more detailed information to be found in the cross-referenced appendices.

Chapter 1 of this book is a general introduction to ALWC. It covers the types of structures at risk, the implications of its occurrence and a brief overview of relevant health and safety and environmental issues.

Chapter 2 describes traditional marine corrosion mechanisms and associated corrosion rate data for steel maritime structures exposed to tidal influences in order to put ALWC into context. It expands on the basic ideas outlined in Chapter 1, providing more detailed information on identification, damage patterns, proposed mechanisms and reported rates of metal loss for ALWC.

Chapter 3 concerns the geographical distribution of ALWC and the surveys carried out to date, including the questionnaire survey, as part of this work.

Chapter 4 introduces the general principles of asset management in a maritime environment with particular emphasis on ALWC.

Chapters 5 and 6 deal with the interrelated subjects of condition appraisal, repair and protection.

Chapters 7 and 8 look at monitoring the performance and procurement of remedial methods relevant to ALWC.

Chapter 9 focuses on areas requiring funding for further research into the problem, its identification and treatment.

Chapter 10 gives guidance on general good practice in dealing with ALWC.

The bibliography and cited references provide a comprehensive list of both technical and commercial literature relevant to ALWC. The glossary of terms will assist the reader in understanding the multidisciplinary terms associated with the phenomenon of ALWC.

What causes ALWC? The detailed mechanism of ALWC is still unclear, but it involves a combination of microbiological and corrosion processes, which may explain the unpredictability of its occurrence.

Where does it occur? ALWC typically occurs just above the area of lowest astronomical tide (LAT) although it has recently been reported as present below this level. The distribution of ALWC is both highly variable (some elements of a structure will have it while their neighbours will not) and consistent (it will often occur at the same location for particular pile or other element geometries, irrespective of geographic location).

Will I get it? No reliable predictive model is currently available for assessing the risk of ALWC appearing at any particular location. Therefore, there is currently no location where the possibility of it occurring can be ruled out.

The extent and spread of the ALWC problem is a source of ongoing debate. Some claim that the problem is exaggerated with only relatively few locations being affected. Others claim that it is already widespread and even if a structure is not currently being affected, it is only a matter of time (or diligence in looking for it) before all experience it, unless appropriate protection measures are taken. Diligence in inspection and maintenance should, therefore, be a priority.

A recent survey, initiated as part of this study, indicated that of the 70 questionnaires returned, 57 per cent stated that ALWC was present while 23 per cent reported no significant corrosion. A more comprehensive and widespread survey of steel maritime structures by experienced personnel familiar with ALWC is required in order to obtain a definitive answer as to the scale of the problem within the UK.

How do I know if I have ALWC? Currently the only reliable method of detecting ALWC is by visual inspection (either by boat at very low spring tides or using divers) together with residual wall thickness measurements. ALWC occurs as localised patches of damage, identified by a characteristic, poorly-adherent orange corrosion product over a 'soupy' black (iron sulfide rich) underlayer associated with rapid metal thinning (the substrate steel surface exhibiting a bright, clean and shiny appearance with extensive pitting). An area exhibiting this visual characteristic alone but without the local high rates of attack is not confirmation of ALWC.

How can I manage ALWC and what will it cost me? The strategy for management of ALWC will depend on whether a structure is new build (where the aim will be to take measures to minimise the risk of ALWC occurring) or an existing structure in the early, intermediate or advanced stage of attack (where the aim will be to take appropriate action to ensure it remains operational for the required remaining service life). Design, inspection, structural analysis, repair, protection and on-going monitoring of maritime structures must now be done taking ALWC into account with due regard to relevant health and safety legislation and environmental responsibilities.

The corrosion protection measures that are currently applicable to ALWC are those based on conventional corrosion control methods which have a track record of protecting steelwork from other forms of corrosion in a marine environment, notably cathodic protection (CP) and coatings of various types. The criteria for optimal protection are not as clearly understood as for 'normal' marine corrosion, but all can be expected to contribute to the protection against ALWC to a greater or lesser degree. The real challenge with ALWC, as with most corrosion problems, is to obtain safe and long-lived service for the lowest overall cost. ALWC only affects a small percentage of total exposed surface area on a structure, and an area for future research is developing remedial techniques aimed specifically at treating ALWC in a more focused way.

In the absence of an accurate predictive model, it is prudent to assume that ALWC will occur at all locations. In general, the adage "a stitch in time saves nine" applies and the sooner corrosion control measures are installed, the lower the overall cost will be. The cost of a combined corrosion protection system, ie CP and coating, generally recommended for new build structures, typically accounts for some two to 10 per cent of overall construction costs. A life cycle cost analysis should be performed to compare the actual financial implications of candidate corrosion control options for particular structures. Installing corrosion control to an existing structure will typically cost three to 10 times as much, while repair of a badly corroded structure will eventually become more expensive than replacement with a new structure. Subsequently the proper maintenance of the corrosion control system is also essential for cost effective performance and should be properly resourced and managed.

As with all significant assets, a whole life costing and asset management approach should be adopted in order to minimise the total cost of ownership of the asset. Taking into account the probability of ALWC, the provision of corrosion control should significantly reduce the overall cost, whether installed at the outset or as a retrofit.

How will it impact on my business? If ignored, ALWC can result in extensive, unbudgeted and costly repair and maintenance works at an unexpectedly early stage in the life of a structure. The damage caused by ALWC is not obvious. It occurs below the low water mark and, unless you look for it, the first sign of the problem is liable to be loss of containment of backfill resulting in the collapse of surfaces above. ALWC in an advanced stage could also potentially lead to the collapse of a structure. By the time holing happens any repair is likely to be expensive, not only in terms of the direct cost of the repair process itself but also, and probably more significantly, in terms of the loss of operational usage. Repairs disrupt operations. In contrast, early identification and remedial treatment will be relatively inexpensive and can often be managed to minimise disruption.

Can I delay spending money? Providing the structure has been analysed by a competent design engineer who has determined an acceptable remaining serviceable life without the need for immediate repair and/or protection, major spending on remedial measures can be deferred. However, this does not mean completely avoiding spending money – regular inspection should be continued, especially on critical elements in the areas subject to ALWC. Furthermore, as the degradation of the structure proceeds, the eventual repair costs will mount. Therefore, regular inspections followed by early remedial measures are recommended. The sooner ALWC is detected and addressed, the cheaper the remedial solutions will be.

How do I go about it? Owners should consider corrosion protection, including from ALWC, on all maritime structures, both new build and existing. They should be fully aware of their technical, commercial and legal responsibilities for managing operational, financial, environmental and health and safety risks. The following recommendations are made:

For new build structures:

- include cathodic protection in any design (it can protect against other forms of corrosion in addition to ALWC. ALWC may cause the loss of up to two thirds of design life if not protected against)
- if CP is not included initially, ensure electrical continuity is provided between all elements of the structure to facilitate its installation if required at a later stage
- apply a heavy duty, protective coating system to at least the exposed part of the structure in the atmospheric, splash and tidal zones and preferably the retained face. (This will reduce the CP current requirement for protection which, in the case of galvanic CP will reduce the number and weights of anodes required or prolong anode life, and will provide protection in the regions where CP is ineffective)
- record all design data including specific minimum thickness requirement at the critical low water zone
- institute a monitoring programme to confirm the efficacy of the protection applied and determine maintenance timing.

For existing structures:

- where necessary, seek independent, professional advice for guidance on the most appropriate and cost effective remedial strategy and when to apply it for the particular structure
- inspect all structures as soon as possible and establish the presence of any locations showing high rates of metal wastage or holing (particularly on critical elements) and pattern of such attack. Confirm the cause and treat appropriately. Note that competent inspection is required for the accurate identification of ALWC
- if ALWC is the cause, identify the generalised (ideally maximum) levels of steel loss. It is important that a consistent approach to measurement be adopted from the outset to allow future comparison. Calculate the rate of loss for forecasting purposes when setting a detailed inspection programme
- for the attack found, re-analyse the factor of safety against overstress, instability or collapse. Note that modern analytical techniques may find useful redundancy, but you should not rely on the structure's apparent ability to withstand collapse, even with holes
- establish a priority work programme. Keep and update regularly a condition database for all parts of all structures and use risk management techniques, as appropriate
- propose a programme of remedial measures and an appropriate budget and calculate potential loss of asset value at collapse or for reduced loading use
- report to Budget-holders/management making clear their Health and Safety responsibility and establishing full board level backing appropriate to the importance of the matter
- carry out priority repairs (holes > excessive thinning > overstress). Make emergency repairs a priority over all but initial inspection
- repair operations can be combined with inspection and save cost. Remobilisation is often a wasted expense. Working in the dry environment of a limpet dam for quality results is often advisable
- do not let holes occur since they complicate the repair work and waste money
- continue to monitor repairs and protection systems.

Remember:

- failure to implement appropriate measures for the monitoring and control of corrosion can expose an organisation and its management to the full severity of Health and Safety and Environmental Legislation if a fatal incident, pollution release or harm to the public occurs
- following and documenting the procedures identified in this guide and carrying out the investigation, design and construction work arising from them will demonstrate that an owner and his technical and commercial management have intended to meet their legal responsibilities for health and safety, and the environment.

Contents

List of figures

List of plates

List of tables

List of boxes

Abbreviations

AC	Alternating electrical Current
ACOP	Approved Code of Practice
ALARP	As Low As Reasonably Practicable
ALWC	Accelerated Low Water Corrosion
APB	Acid Producing Bacteria
ASTM	American Society for Testing Materials
CD	Chart Datum
CDM	Construction (Design and Management) Regulations 1994
CP	Cathodic Protection
DC	Direct electrical Current
DFT	Dry Film Thickness
HAZ	Heat Affected Zone
H&S	Health and Safety
HSWA	Health and Safety at Work Act (1974)
ICCP	Impressed Current Cathodic Protection
ICE	The Institution of Civil Engineers
ILW	Intertidal Low Water
ISO	International Organisation for Standardisation
LAT	Lowest Astronomical Tide
MHWS	Mean High Water Spring
MHWN	Mean High Water Neap
MIC	Microbiologically Influenced Corrosion
MLWS	Mean Low Water Spring
MLWN	Mean Low Water Neap
NACE	National Association of Corrosion Engineers, USA
NLWC	Normal Low Water Corrosion
O&M	Operating and Maintenance
PWC	Preferential Weld Corrosion
QRA	Quantitative Risk Assessment
QUENSH	Quality, Environment, Safety and Health
SCE	Saturated Calomel reference Electrode
SOB	Sulfur Oxidising Bacteria
SRB	Sulfate Reducing Bacteria
SLS	Serviceability Limit State
ULS	Ultimate Limit State
VOC	Volatile Organic Content

1 Introduction

1.1 Background

ALWC (accelerated low water corrosion), sometimes also referred to as "LAT (lowest astronomical tide) corrosion", is a particularly aggressive form of localised corrosion defined in the British Standard for Maritime Structures (BS 6349-1: 2000) as a type of low water "concentrated corrosion". It has become a high profile problem, associated with unusually high rates of metal wastage and holing (see Plate 1.1) on unprotected, or inadequately protected, steel maritime structures. Average corrosion rates in the range of 0.3 to 1.0 mm/wetted side/year are typically reported but higher instantaneous rates are probable once ALWC has initiated on a structure. If timely action is not taken, this rapid form of attack can result in extensive, unbudgeted and costly remedial repair and maintenance works at an unexpectedly early stage in the life of a structure. ALWC in an advanced stage could also potentially lead to the collapse of a structure. It is not surprising, therefore, that the occurrence of ALWC was described as "a matter of national importance" at a conference organised by the ICE on the subject in London, October 1998.

ALWC can occur in tidal waters on steel maritime structures and is most commonly found just above LAT, although it has been reported at areas below this level. The level at which it occurs means that, unless efforts are made to survey the structure in the susceptible zones, it can remain undetected. ALWC has a recognisable appearance (see Plate 1.2), typically showing patches of lightly adherent, bright orange and black (iron sulfide rich) deposits over a clean, shiny and pitted steel surface. As the pits deepen, become more numerous and overlap, this produces a dishing effect in the metal surface, which ultimately develops into a hole.

The detailed mechanism of ALWC continues to be a matter of some debate but the evidence (see Chapter 2, Section 2.3.3 and Appendix 3) is that ALWC is associated with bacterial activity and it is, therefore, a form of microbially influenced corrosion (MIC). Although localised (or accelerated) corrosion on steel maritime structures at low water can be caused by other mechanisms, a defining characteristic of ALWC is that it usually conforms to a specific pattern of damage on steel sheet piled structures. While attack is often seen to be random both within and between installations, the pattern of damage is usually similar for particular pile geometries, irrespective of the geographic location. The presence of poorly adherent orange and black (iron sulfide rich) patches of surface deposit and local high rates of metal loss occurring together often confirms ALWC but not if one of these features is absent (see Chapter 2, Section 2.3.1).

There has been some confusion in defining ALWC in the past, both in literature and from the maritime industry. Some have rightly acknowledged it as a discrete corrosion phenomenon and a form of MIC. Others such as BS 6349-1: 2000, have referred to it collectively, with other corrosion mechanisms that can give high rates of attack in the low water and submerged zones, as "concentrated corrosion". However, others have used the latter term to define the problem in the low water zone specifically in terms of macro concentration cells, principally differences in oxygen concentration, with no acknowledgement to the role of MIC (OCDI 1998). Even a recently published maritime inspection guidance document (PIANC, 2004), which acknowledged the possible contribution of MIC to low water corrosion, does not refer to the ALWC phenomenon.

An early reference to the existence of black (iron sulfide rich) deposits, indicative of MIC, occurring on corroded steel structures immersed in seawater appears in a South African technical paper (Copenhagen 1934). A subsequent paper by the same author twenty years later (Copenhagen 1954) describes a specific low water corrosion problem on steel sheet piling at Cape Town Docks. Corrosion rates of 1.25 mm/side/year were observed, associated with black, sulfide rich deposits reported to be "due to the action of sulfate reducing microorganisms". The author at that time stated that "this underwater type of marine corrosion is far more common than has been hitherto suspected" and considered the problem "to be of sufficient importance to examine bottom muds of harbours from different parts of the world for sulfide content". These included samples taken from English ports (Port of London and Southampton Docks). However, this would appear not to be the first recorded incident of what is now commonly – although not universally – referred to as ALWC. An identical or very similar incident was reported occurring on steel-joist piles in Hong Kong harbour back in 1914 (Ellis, 1914). The problem was described by the author at the time as occurring "in a rather curious and alarming manner" being restricted to surfaces at and below the low water zone. Pits and "saucer shaped depressions" filled with black iron sulfide were found on relatively new structures, associated with corrosion rates estimated at 2 mm/ wetted side/year. The generation of hydrogen sulfide by "certain marine growths" was cited by the author as a possible causative agent.

According to more recent literature, the existence of ALWC is first reported as being identified in the UK in the early 1960s (Morley, 1978) but was not widely recognised as being present in UK waters until the 1980s (BS 6349-1, 2000). In a survey of UK ports and harbours carried out in 1980 (Doughty, 1980) no unusual corrosion behaviour was reported in the low water zone. While the authors suspected the possibility of MIC occurring in ports and harbours at this time, little evidence for this had been forthcoming from the survey itself. Since then, however, the number of reported cases of ALWC on maritime steel structures appears to have increased significantly both within the UK and worldwide. It is now recognised that such localised (accelerated) corrosion, by whatever means, does unquestionably occur on steel maritime structures at or around LAT.

Consequently, to minimise the risks of failure of marine structures, infrastructure owners, operators and engineers should now plan design, inspection and maintenance taking into account the possibility of ALWC (and any other localised corrosion mechanism) occurring in the susceptible zones. Ignorance would be no defence against any charges of negligence for structures failing from ALWC, or other localised corrosion attack, through lack of maintenance or deficient design provision.

See Plate 1 *ALWC holing at low water on typical maritime structure* and Plate 2 *Close up of typical ALWC patch* on page 98.

1.2 Purpose and scope of work

This guide is aimed at three main user groups:

- infrastructure owners and operators, and their advisors
- specialist engineers – experts in marine activities
- asset and maintenance managers.

The book addresses the technical issues associated with ALWC in design, condition appraisal, repair, maintenance and protection. It promotes managerial and engineering best practice to reduce the impact of ALWC on steel maritime structures. It is based on:

- a detailed review of published technical and commercial literature
- consultation with experts and practitioners within a cross-section of fields of expertise
- information obtained from a questionnaire-based survey
- case studies demonstrating good practice.

The purpose of the book is to:

- provide an independent overview of ALWC and associated problems – both nationally and globally
- increase awareness and enable early identification and repair of ALWC, thereby permitting early action to minimise health and safety risks to users of maritime structures
- improve protection/reduce risk of failure of steel marine structures
- recommend appropriate design, maintenance and repair strategies in order to reduce whole life costs
- combine previous research with current good practice in one document.

Some chapters will have more relevance to some practitioners than to others, however all will gain an insight into the factors that govern the management of ALWC by reading the whole book.

1.3 Applications and types of structure at risk

Any unprotected, or inadequately protected, steel structure exposed to tidal or impounded marine or brackish environment is susceptible to ALWC which can occur on any wetted surface of any pile type and/or fittings, such as ladders, chains, fenders, etc. This document is applicable to the following types of structure at risk from the effects of ALWC:

- Skeletal structures, with steel tubular, box and H etc pile sections:
  - jetties
  - piers
  - dolphins
  - wharves.

- Solid structures, with steel retaining structures, including:
  - wharves
  - quays
  - cellular cofferdams
  - retaining walls
  - canal and river walls in estuarine areas, including bank protection structures
  - dolphins
  - lock gates and sluices
  - dams
  - coast protection works
  - bulkheads
  - breakwaters
  - flood defence structures
  - barrier walls in cooling water system intakes/outlets.

Steel sheet piling is widely used in the construction of many types of maritime structures and there are, therefore, many kilometres of sheet piling potentially at risk from ALWC attack. Steel piling was introduced as a successor to cast iron at the beginning of the 20th century. Steel sheet piling consists of hot-rolled or cold formed steel sections that can be interlocked to produce structures ranging from a simple continuous corrugated wall (where individual pieces are driven side by side) to "combi walls" made up of king piles and intermediate sections. The Universal pile system (based on connected universal joists) was created in 1907 with a lighter version, the Simplex system, following in 1912. The types commonly used today (see Figure 1.1) were first introduced in the UK in the 1920s (ie U piles) and in the late 1930s (ie Z piles).

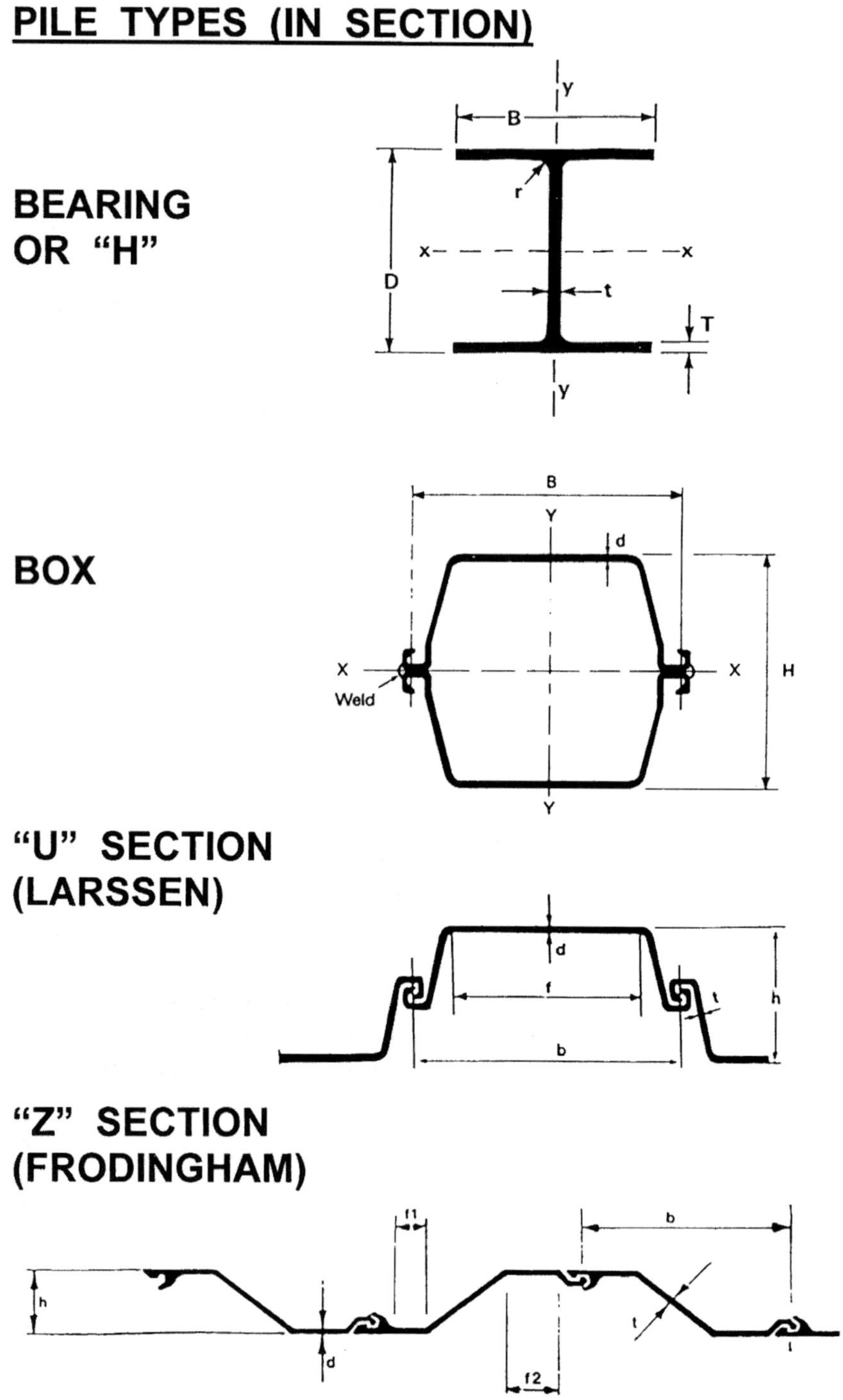

Figure 1.1 Pile types commonly used in maritime construction

Other unprotected steel members also at risk from ALWC in the low water zone include:

- steel repair plate
- angle members
- mooring chains
- ladders
- steel elements of fendering systems.

1.4 Implications of ALWC to structures

ALWC produces rapid, local metal thinning which can, if left unchecked develop into:

- serious holing and the need for urgent repairs
- premature structural failure leading to partial or complete reconstruction of a structure or even total shutdown of a facility.

Such features have been observed after a service life of as little as 20 years on structures that were originally designed to give 40 to 120 years. Clearly structures designed with thinner steel sections could be at greater risk. However, it is not yet possible to accurately predict if a structure will suffer the problem or at what stage in the life of a structure the problem will initiate (Gubner, 1998).

The forms of failure caused by ALWC are:

- **Loss of containment or stability**; BS6349-1:2000 warns that loss of backfill material through a holed sheet pile wall can lead to the collapse of pavement surfacing above or structures supported on the backfill. Uneven surfaces in themselves can also prove hazardous. In the case of ports and harbours, significant loss of fill material can, in extreme circumstances, reduce the navigable depth of a berth creating a potential hazard to transport, cargo and maintenance services.

  Loss of containment producing voids behind a steel sheet piled wall will become apparent during inspection, either planned or following the failure of the adjacent supported pavement. The impending loss of containment will be indicated by local thinning or initial holing of steel forming the retaining structure. Often local clouding of the water on a falling tide caused by loss of silty material can warn of a hole.

See Plate 3 *Holed Z section pile showing loss of backfill material* on page 99.

- **Loss of strength or structural failure**; loss of structural section can weaken the affected structure, potentially leading to failure or the need to place loading restrictions on it. Although ALWC only affects a small percentage of total exposed surface area, it has potentially serious implications to structures where the resultant loss in section from ALWC corresponds to areas of maximum stress. Structural failure has fortunately been rare and sometimes does not occur even when detailed analysis suggests that it should have done so. Nevertheless, it is essential for the owner to be confident that a structure is operating safely and can carry the loads that will be imposed on it.

  A potential reduction in the load carrying ability of the structure may be identified by loss of section and/or holing. The structure should be analysed by a competent designer, either:

  - to prove the adequacy of the structure and assist in determining the remaining serviceable life without protection
  - to determine the extent of necessary repairs and protection strategies.

Simply replacing the backfill before carrying out remedial works may cause over-stress of a steel sheet piled wall already weakened by ALWC. Allowing the structure to hole before suitable remedial works are performed often complicates and increases the cost of carrying out such repair works.

1.5 Brief overview of Health and Safety legislation

All activities related to ALWC, in common with all workplace activities, are covered by the Health & Safety at Work Act 1974. In almost all cases – except perhaps the most basic inspection and survey work, such as "taking levels, making measurements and examining a structure for "faults" – inspections, surveys, repair and protection work associated with ALWC affected structures and new build will be covered by CDM Regulations (HMSO, 1994). However, even with basic inspection and survey work, the addition of any significant work to clean off marine biofouling from surfaces would appear to bring the work into the scope of the regulations. It would, therefore, be wise in view of the potential risks to follow guidance given in the regulations.

The Approved Code of Practice (ACOP) "*Managing Health and Safety in Construction: Construction (Design and Management) Regulations 1994*", HSG224 (Health and Safety Commission, 2001), issued by the Health and Safety Executive provides advice and guidance on the application of the Regulations which encompass building, civil engineering and engineering construction work, including cleaning, maintenance, repair, renovation and demolition.

The CIRIA publication "*CDM Regulations – Practical guidance for clients and clients' agents*" (CIRIA C602) provides information on when and where CDM applies and the roles and responsibilities of parties under CDM. The associated CIRIA publication "*CDM Regulations – work sector guidance for designers*" (CIRIA C604) provides general guidance to designers on the identification of hazards in relation to the health and safety of construction workers and those affected by construction work. It is important for all parties to understand their responsibilities under CDM and, in particular, to recognise that the term "designer" can encompass a number of activities including temporary works and surveys.

There are many potential hazards associated with repair and protection work for structures affected by ALWC that may need to be considered. The following list gives examples but is not exhaustive:

- ensuring stability of the weakened structure at all stages of the work
- working under, over and adjacent to water
- interaction with facility operations – shipping and road/rail traffic
- working at height
- blasting and coating work within a confined area
- lifting of heavy or awkward items (eg anodes, steel repair plates, etc)
- use of hazardous materials with potential to cause harm (eg fuel, cement, organic solvents, biocidal washes, etc)
- use of ultra high pressure water jetting
- leptospirosis.

Designers should give particular consideration to the means of safe access to undertake future inspection and maintenance on their structures, particularly where these features (eg access ladders) can be included as part of the construction work or added during repairs.

1.6 Environmental responsibilities

Environmental legal requirements will vary between countries but will almost certainly have an impact on the:

- design and form of construction of the structure
- materials incorporated in the permanent works
- control of the process by-products
- selection and use of plant and equipment.

There are many potential environmental hazards associated with protection systems for new build structures and repair and protection work for structures affected by ALWC which need to be considered. The following list gives examples but is not exhaustive:

- coating systems (see Appendix 10, Section 1.1)
- jet/blast cleaning activities (see Appendix 10, Section 1.4)
- proper waste disposal from all works
- galvanic anodes for cathodic protection of structures at locations where there are already high concentrations of heavy metals in the seawater from whatever source (see Appendix 10, Section 2.0).

In many cases, environmental impact studies may be required and licences first obtained before work can commence, usually under strict terms of the authorities involved. In general, the repair and protection work associated with ALWC will be subject to the same environmental controls as any other similar work at that particular location.

Within the UK, the extent of the statutory controls will depend on the location, extent and magnitude of the work. The legislation given below may apply in addition to any requirement for a Harbour Works Licence. Application is made to the Marine Consents and Environment Unit (MCEU), a central facility for receipt and administration of applications to undertake works in tidal waters below MHWS and at sea. In considering applications the MCEU will consult with various governing bodies, agencies, and departments, including those shown below.

- **Coast Protection Act 1949**

  Imposes restrictions on works which may be detrimental to the safety of navigation or have potential environmental effects.

  Consultees: Environment Agency, English Nature, Countryside Council for Wales, Crown Estate, English Heritage, Local Authority via the planning department, Sea Fisheries Committees and competent navigation authorities.

- **Harbour Works (Environmental Impact Assessment) Regulations 1999**

  Applies to Coast Protection Act applications involving harbour works.

  Consultees: English Nature, Countryside Council for Wales, Environment Agency, Countryside Agency, DEFRA and local authorities.

- **Food and Environment Protection Act (FEPA) 1985**

  Controls the deposit of articles or materials in the sea or tidal waters and primarily aims to protect the marine ecosystem, human health, and minimise interference to others by considering any potential hydrological effects, possible interference with marine activities, turbidity, noise, drift of fine materials to smother seabed flora and fauna and adverse implications for designated conservation areas. In Scotland, an FEPA Licence is obtained from the Scottish Executive.

Consultees: Government agencies and departments with navigational interests, maritime and coastguard agency, Corporation of Trinity House, SEPA, local harbour authorities, fishermen's organisations and leisure interests.

The following statutory controls may also apply to work in, or adjacent to, protected or conservation areas:

- **Conservation (Natural Habitats &c.) Regulations 1994**

  If the work is in, or adjacent to, a European Site Special Protected Area (SPA) or candidate SPA, or Special Area of Conservation (SAC) or candidate SAC the MCEU may also consult with English Nature, the Countryside Council of Wales and the Centre for Environment, Fisheries and Aquaculture Science (CEFAS) to determine whether it is likely to have a significant effect, and to determine any mitigation measures to be put in place.

- **Countryside and Rights of Way Act 2000**

  If the work is in or adjacent to a Site of Special Scientific Interest (SSSI) the MCEU may also consult with English Nature to determine whether the works are likely to potentially affect the species or features protected by the site and to determine necessary conditions to be included within any consent.

Depending on the specific location and level of the work, other more specific acts such as the Land Drainage Act 1991 may also apply.

Certain remedial works and operations may be exempt from particular requirements. For example, exemption from the Food and Environment Protection Act may be granted where remedial works are to refurbish or replace parts of existing structures or facilities with materials or articles of a similar nature (essentially "like-for-like" replacement) to extend the serviceability of the structure or continue its purpose. However, this will not apply if the remedial works will significantly enhance, alter, or extend the existing structures or facilities, or if the remedial works require additional temporary works to be constructed adjacent to the existing site.

CIRIA have produced two guidance documents providing good practice guidelines for construction and maintenance work.

- "*Coastal and marine environmental site guide*" (C584) published in 2003 addresses issues in the marine environment
- "*Environmental good practice on site handbook*" (C502) published in 1999 addresses issues in the freshwater and terrestrial environments.

Summary

- ALWC is a high profile problem that can produce rapid rates of localised attack on steel maritime structures, typically just above LAT. It is a discrete corrosion phenomenon and a form of MIC (see Chapter 2)
- if ALWC is left unchecked, the consequences are loss of stability and/or gradual weakening of the affected structure potentially leading to structural failure
- any unprotected, or inadequately protected, steel structure (eg steel sheet piled walls, tubular piles and so on) exposed to tidal marine environment is susceptible to ALWC
- since the 1980s the number of reported cases of ALWC on steel, maritime, structures have notably increased both within the UK and worldwide (see Chapter 3)
- this book addresses the management of ALWC to reduce its impact on maritime structures (see Chapter 4) and highlights the technical issues. It proposes methodologies for condition appraisals (Chapter 5), repair and protection for new build and existing structures (Chapter 6), monitoring (Chapter 7) and procurement of remedial methods relevant to ALWC (Chapter 8). Areas requiring funding for further research into the problem, its identification and treatment are presented in Chapter 9. Finally, guidance on general good practice in dealing with ALWC is given in Chapter 10.

2 Background to ALWC

This chapter provides a brief introduction to:

- "traditional" (ie oxygen diffusion controlled) corrosion behaviour of steel under the varying conditions of immersion in unpolluted seawater
- the forms of corrosion that are capable of producing high rates of localised (accelerated) attack on unprotected steel in the low water zone to put ALWC into context
- the characteristics of ALWC and the mechanisms that have been proposed in the literature to explain its action
- published corrosion rates for steel in a marine environment subject to tidal influences including those reported for ALWC.

A more detailed review of the corrosion of carbon steel in seawater is given in Appendices 1 and 2. In addition a bibliography is provided at the end of this book which lists papers, books and standards on steel corrosion and protection in marine environments with particular relevance to ALWC. While all references concerning ALWC have been included for completeness, it should be recognised that the content and advice given in some of these has been superseded as the knowledge of ALWC has been (and continues to be) developed.

2.1 Traditional marine corrosion

Carbon steel is the basic material used for the construction of most floating or fixed maritime structures. Its corrosion behaviour in unpolluted seawater is generally controlled by the concentration of dissolved oxygen available at the steel surface. This produces the customary, usually adherent, hydrated iron oxide deposits on exposed steel, commonly known as "rust". The rate of corrosion is usually governed by the rate of diffusion of oxygen through the layers of rust and/or marine biofouling that build up on unprotected surfaces over time. The rate of such oxygen diffusion to (and hence corrosion at) the surface tends to reduce with time reaching a "steady state" value as the thickness of these surface deposits increases. Depending on the conditions, "rust" can take a variety of physical and chemical forms, differing in colour, hydration and bulk. For example, it is common to find a layered corrosion product scale (see Figure 2.1) consisting of:

- an inner layer of black magnetite (Fe3O4) formed under limited oxygen conditions
- a thin layer of its green hydrate
- an outer coating of ordinary, yellow-brown "rust" (Fe2O3.H2O) formed due to unrestricted oxygen access.

The corrosion resistance to seawater is much the same for all carbon steels (both mild steels and higher strength steels). It might, therefore, be expected that all steel maritime structures of similar geometry seeing the same environment and circumstances of usage would corrode at a similar rate and show similar metal wastage patterns on the structures. However, some variation may be expected on particular structures due to factors such as:

- steel composition – where different grades of steel are in contact
- surface condition, such as distribution of attached marine biofouling, etc
- nature of surface deposits, including rust, iron sulfide, and so on
- local variation in exposure conditions – in salinity due to mixing of seawater with river water, tide levels, temperature, flow, erosion, etc
- vibration
- local and/or repeated stresses
- mechanical working, such as bending of steel plate.

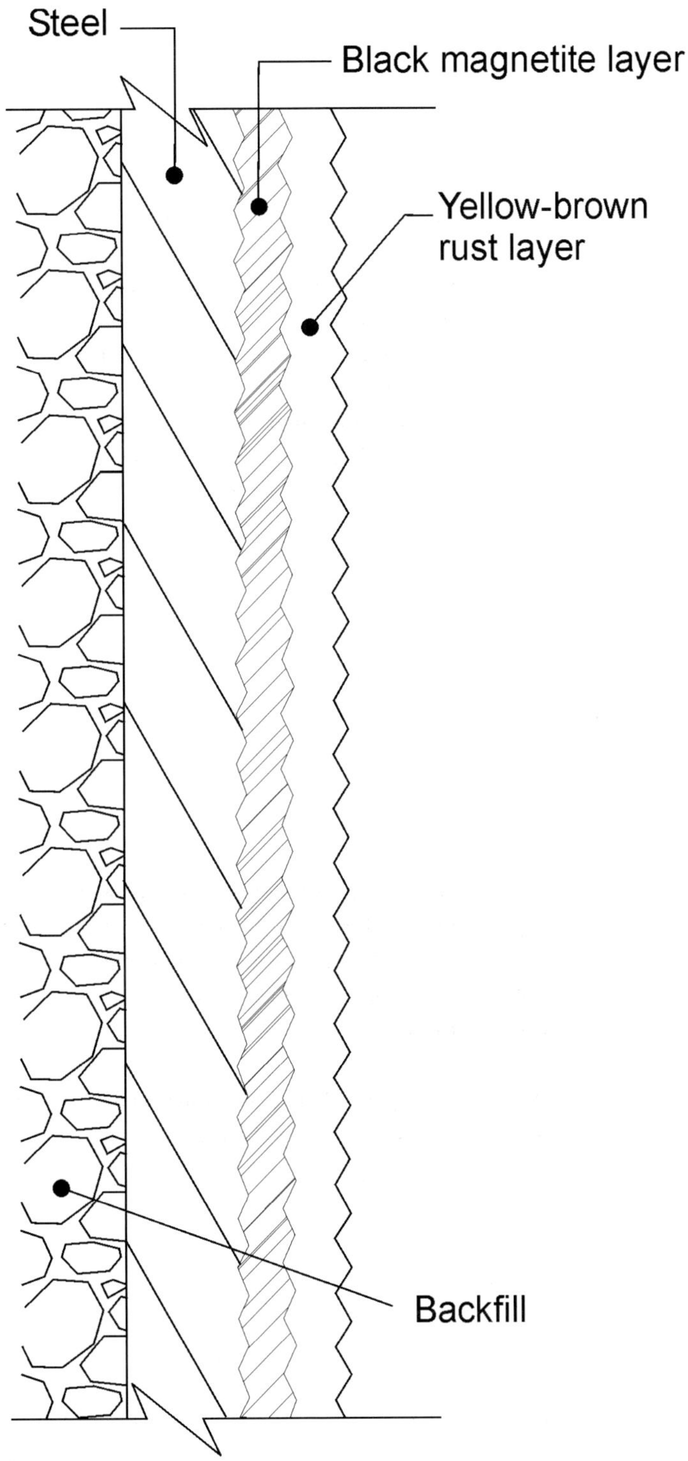

Figure 2.1 **Section through "traditional" marine corrosion product on steel substrate**

Considerable variations can also result from the type and "geometry" of a maritime structure such as the configuration of member cross-sections; for example, horizontal surfaces tend to remain wet longer than vertical surfaces. Physical changes (eg cold working) or heat treatment to a metal may also affect susceptibility to corrosion (Melchers, 1994).

Figure 2.2 (eg INCO, 1973; Humble, 1949) shows a typical corrosion profile for unprotected steel in unpolluted seawater and represents the varying conditions of immersion which the steel may encounter when used in marine applications due to tidal influences. The corrosion rates are averaged, general (uniform) "steady state" values and have been obtained from measurements taken in several world-wide studies carried out over periods of up to 50 years. Much of this information will have been derived from test panels or by measurements on pier supports or piling in maritime environments and is recorded in several books and published papers. In practice, some variation in actual corrosion rates would be expected from structure to structure (Phull, 1997) and, although the pattern of corrosion will generally remain the same, the curve may be displaced to either the left or right depending on local conditions. Some variation in excess of these "steady state" corrosion rate values may also be possible during the initial stages of exposure.

It is normal to divide the marine pile into six vertical zones, namely:

1. Atmospheric zone.
2. Splash zone.
3. Tidal zone.
4. Intertidal low water (ILW) zone.
5. Continuous immersion zone.
6. Mud (embedded) zone.

The corrosion mechanisms operating in the six exposure zones have been well documented in the literature, some zones being more vulnerable to attack than others. Examination of the typical corrosion profile (Figure 2.2) reveals that the main zones of attack are traditionally seen in the splash zone and around the low water level. Sometimes a third peak also occurs in the area of the mud line (or sea bed) due to the scouring action of sand/gravel.

A description of each zone and the range of corrosion rates characteristic to each zone taken from the literature are given in Table 2.1 (see Section 2.4).

2.2 Common forms of localised corrosion

High rates of attack on immersed steel structures in the low water and submerged zones can potentially result from one or more of several common forms of corrosion, collectively described as "concentrated" corrosion in BS 6349-1: 2000, Section 59.2.2. These and the other common forms include:

- differential concentration cells; almost any variation in conditions of exposure at the steel surface can produce a corrosion current within a macro-cell leading to localised corrosion. This mechanism, principally the formation of an oxygen concentration cell, is referred to by Japanese researchers (eg OCDI, 1998; Tsurata, 1973) as "concentrated corrosion". This should not be confused with the more general BS 6349 definition of this term

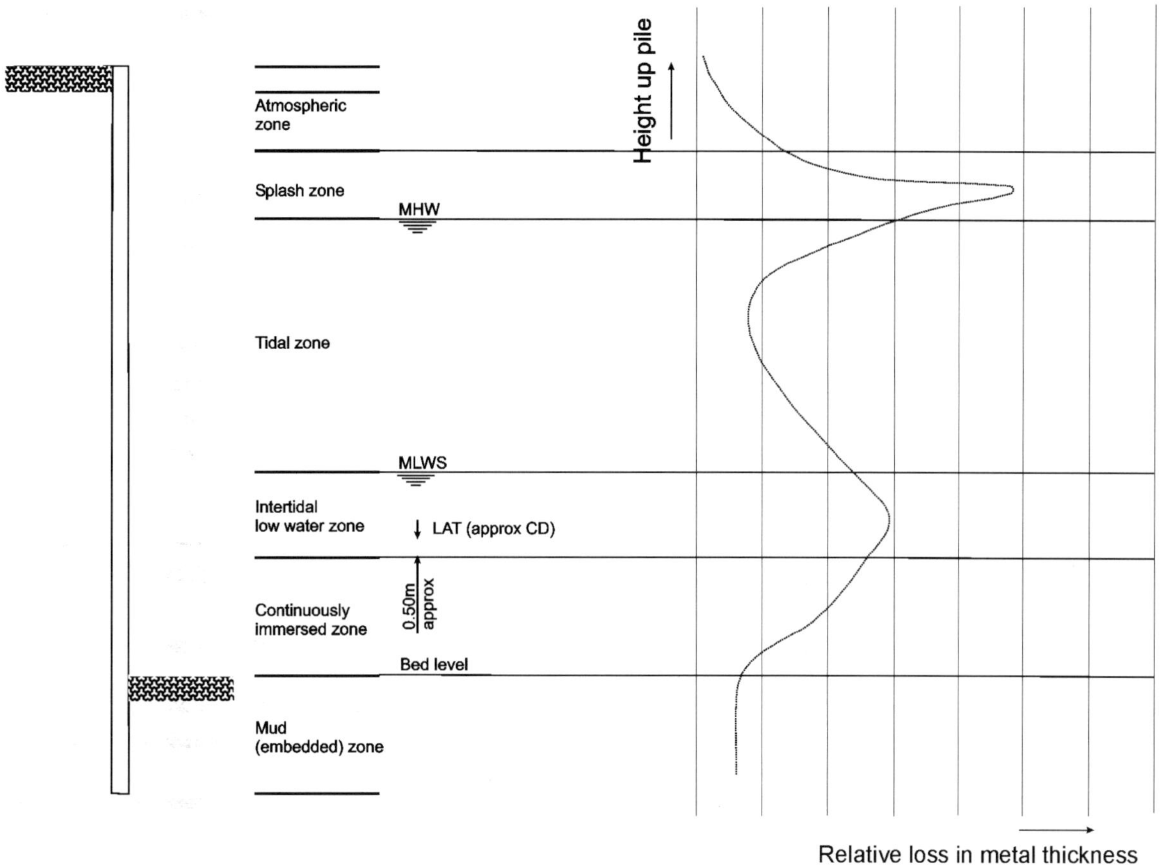

Figure 2.2 **Schematic showing corrosion zones and typical profile for "traditional" corrosion behaviour**

- differential aeration; a specific example of a concentration cell arising from local differences in the availability of dissolved oxygen. Such differences can occur over short distances (eg under marine organisms) or large distances, such as tidal versus low water zone of a steel sheet piled wall. The rate of metal loss can be aggravated where removal of any semi-protective surface deposits also results in the formation of a differential aeration cell
- crevice corrosion; an example of an oxygen concentration cell occurring over short distances, around a narrow aperture or clearance formed between the steel surface and another surface. The latter can either be metallic (eg at mechanical fixings) or non-metallic (eg beneath fouling deposits)
- bimetallic corrosion; typically may occur when two different metals (including steels of differing composition) are connected together and immersed in an electrolyte such as seawater. When it does occur, one metal (the anode) will suffer increased corrosion whereas corrosion on the other (the cathode) will be decreased. The corrosion behaviour will be determined by the electrode potentials of the two metals, the relative areas of the two metals and the conductivity of the environment (for further guidance see Appendix 1 and BS PD 6484). In the absence of suitable corrosion protection, new carbon steel invariably corrodes more when coupled with old carbon steel. Care should also be taken in the choice of consumables when welding onto a marine structure to reduce the risk of preferential weld corrosion (PWC), a form of bimetallic corrosion. A greater intensification of galvanic effects should be expected for metals under immersed seawater conditions compared with

those under marine atmospheric conditions. In the marine atmosphere the bimetallic corrosion tends to be concentrated in the area at or around the point where the metals touch and is largely independent of the relative areas of the anodic and cathodic material

- erosion corrosion; occurs when the velocity of seawater over the steel surface (due to wave action, river currents, tidal currents, ships propellers, bow-thrusters, etc) is sufficiently fast to periodically or continually remove otherwise semi-protective surface deposits. Local metal loss can be significant where the water includes entrained abrasive particles (eg at or near the mud line due to scouring action by abrasive sand/gravel) or where mechanical abrasion occurs (eg due to the repeated, rubbing action of floating fenders against a steel sheet piled wall – Holleywood J, 1978). Adherent surface deposits would have a tendency to reduce the effect of seawater velocity on general (uniform) corrosion although localised attack could be promoted if they were not continuous
- stray current corrosion; occurs when positive current flows through electrolyte paths other than the intended circuits resulting in protection where it enters the steel structure and potentially high rates of local metal loss where it leaves. Possible sources of stray current include direct current (DC) from improperly grounded welding generators, ship service systems, electric crane installations or nearby impressed current cathodic protection (CP) systems (on sections not electrically bonded to the protected structure) or electric railway systems
- microbially induced corrosion (MIC); corrosion caused or accelerated by microbiological organisms on a steel surface, particularly under conditions of stagnation or restricted water movement where bacteria can flourish. In the vast majority of cases, it has been established that the MIC mechanism of attack is to develop a unique inorganic environment close to the metal surface that is highly aggressive. Thus, the microbial role is largely in developing and maintaining an aggressive environment rather than in "eating the metal" directly. The most important group of micro-organisms associated with rapid and severe localised corrosion on steel in marine environments is the sulfate reducing bacteria (SRB) (Lee, 1995). These are anaerobic bacteria and are often found operating in close association with sulfur oxidising bacteria (SOB) and acid producing bacteria (APB), supporting each other's growth. The result of components that promote microbial growth coming together to create a colony on a surface is referred to as a "biofilm". SRB corrosion is characterised by the formation of black, iron sulfide corrosion products on shiny, pitted steel surfaces. Iron sulfides can easily oxidise on exposure to oxygen (air), the black colour often disappearing within hours. The iron sulfides typically oxidise to hydrated iron oxide (ie yellow/brown "rust") and elemental sulfur or sulfates.

2.3 Accelerated Low Water Corrosion (ALWC)

Since the early 1980s, local metal thinning and holing has increasingly been observed on unprotected, maritime, steel structures usually just above LAT. There is no doubt that this has resulted from corrosion which appears to be significantly more severe, both in terms of occurrence and rate of attack, than has previously been reported for this zone under "traditional" corrosion conditions. This "new" phenomenon, associated with bacterial activity, is referred to as ALWC. The location of the problem is rarely exposed by tides and is only completely exposed on affected steel structures at extreme low spring tides, which may only occur four or five times a year.

2.3.1 Identification

ALWC is seen as patches, or a succession of patches, typically in a narrow horizontal band of attack in the critical corrosion zone. On some structures such patches have also occasionally been reported down to bed level (Shelley, 2001). These ALWC patches are identifiable by their characteristic appearance, pattern of damage and high rates of metal loss (see Figure 2.3).

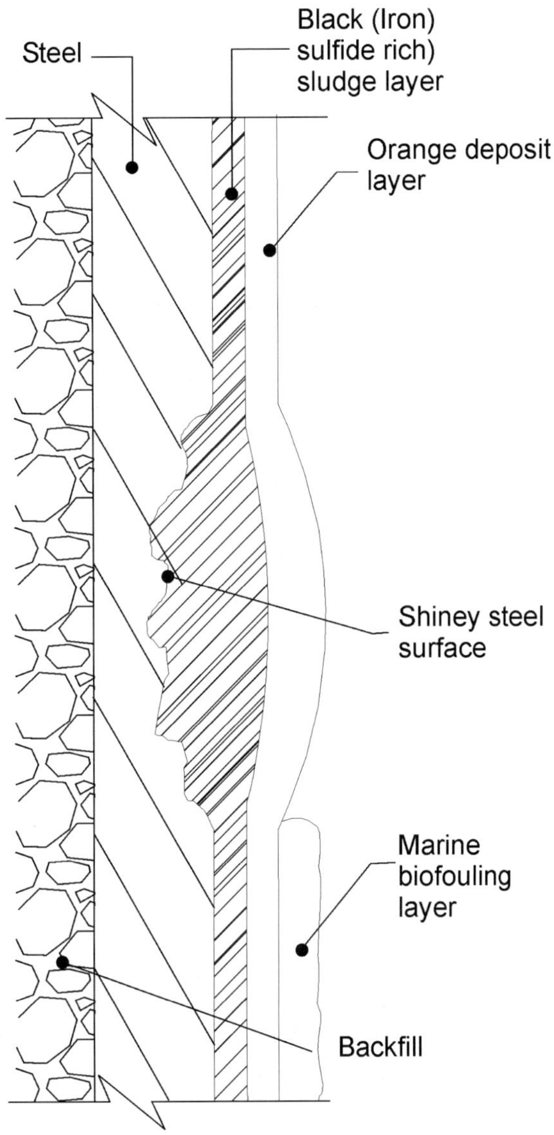

Figure 2.3 **Section through ALWC patch on steel substrate**

(1) Appearance

- *Surface deposit*
    - a two layered structure typically comprising (1) a soft, spongy, bright orange outer layer over (2) a thin, slimy, gelatinous, inner black liquid/paste layer (ie an iron sulfide rich layer)

  Note that:

  1. The surface deposit at ALWC patches is less well adhered to the substrate steel and may be more easy to remove than surface deposit at surrounding areas within the low water zone. This may

also show a similar layered structure to that seen at ALWC patches.

2. The layered surface deposits in the low water zone may occasionally be masked by an outer marine biofouling layer such as algae, mussels, barnacles or seaweed.

- *Substrate steel*
  - bright, clean and shiny surface finish (indicative of active corrosion) with extensive pitting compared with surrounding areas in the low water zone, where the steel is "dull" (ie a matt finish) and often uniformly attacked
  - as pits deepen, become more numerous and overlap, this produces a dishing effect in the metal surface which ultimately develops into a hole

(2) Pattern of damage

In most cases the ALWC patches will be seen to occur randomly both within and between sites. However, a defining feature of ALWC is that the pattern of damage is generally similar for particular pile geometries, irrespective of the geographic location of the site.

- at ALWC affected sites there is no perceived pattern or trend regarding distribution of ALWC. Some sites are heavily affected by the problem while others appear to be unaffected, even those within close proximity to each other and where exposure conditions would be expected to be similar. Reported examples include (1) a harbour where sheet piling on one site showed a heavy incidence of ALWC while those on another site showed little evidence of ALWC (Cheung, 1994) and (2) a series of adjacent quay structures where four quays were found to have the problem but one did not (Shelley, 1998). When it does occur, it may be in clusters along the length of the structure. In the case of a steel sheet piled wall, a pile which is sound may be next to a pile which is not.
- in European ports, supported by evidence from Japan (eg OCDI, 1998) and the USA (Kennedy, 1967), there is a consistent and specific difference in the pattern of damage evident between U- and Z-shaped piles (see Figure 2.4) suggesting a common cause of attack is occurring at all affected sites (Johnson, 1997). The damage to Z piles result in vertical, elongated slits (typically up to 1000 mm in length and 100 mm wide) towards the corner or webs of the outpans. On U-shaped piles and straight web piles, round holes (typically up to 300 mm diameter) occur in the centre of the outpans. It is unclear why the attack occurs on the thinnest part of the Z pile and the thickest part of the U pile although, in both cases, it is the area furthest from the clutch. Holing nearly always appears on the outpans. Inpans are almost invariably unaffected. The problem is also seen on other pile types eg tubular piles, box and H-piles where damage usually occurs on the outward, seaward facing surface.

(3) Rates of metal loss

Average corrosion rates in the range of 0.3 to 1 mm/wetted side/year are typically reported but higher instantaneous rates are probable once ALWC has initiated on a structure. Such localised, high rates of metal wastage at ALWC patches make holing possible within a matter of a few years on most affected structures (see Section 2.4).

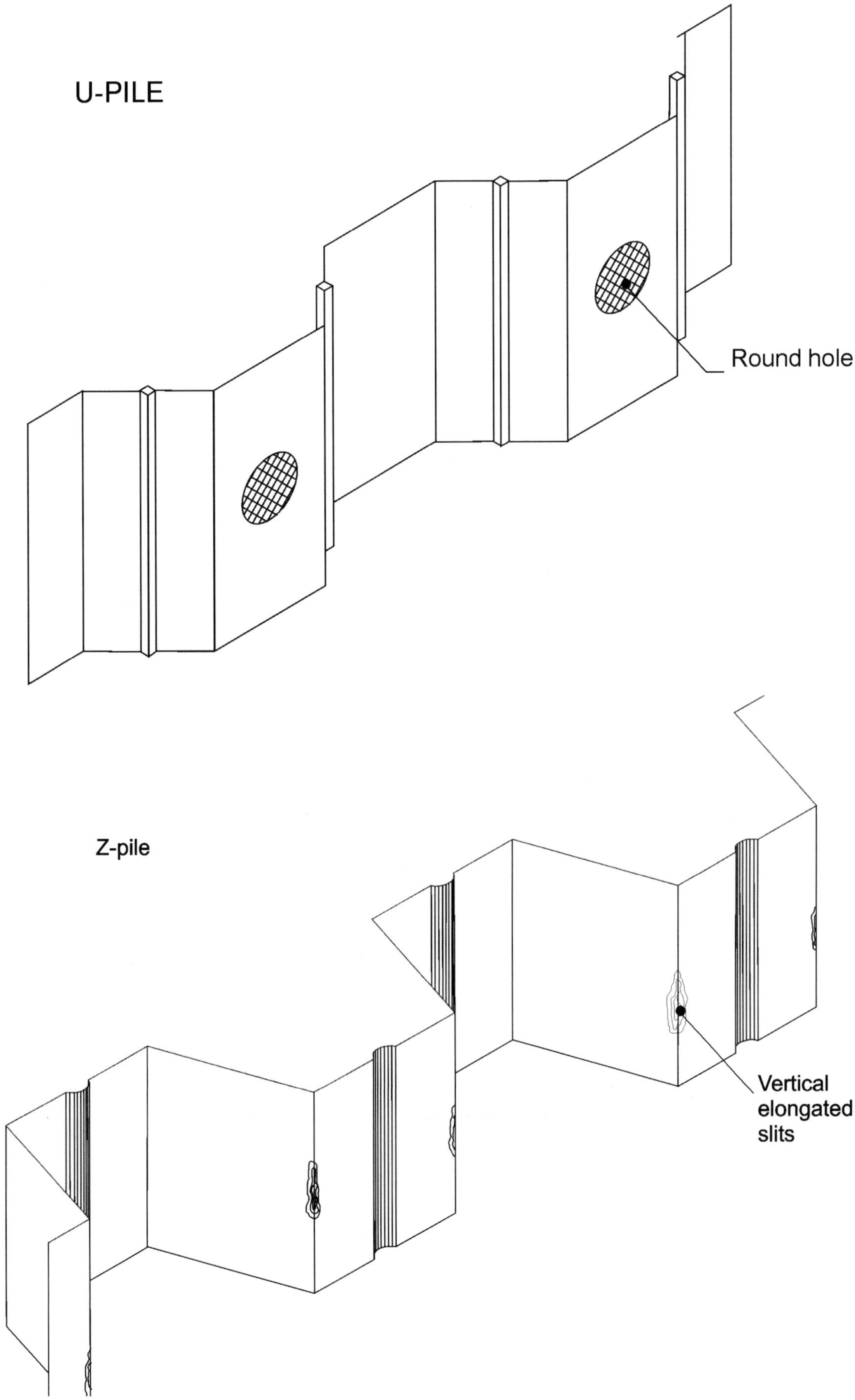

Figure 2.4 Typical patterns of ALWC damage seen on U and Z piles

2.3.2 Influencing factors

Some of the potential risk factors which have been reported to influence the incidence of ALWC, and need to be considered to indicate a predisposition towards ALWC occurring for all steel maritime structures, are presented below. However, their importance in the development of ALWC is difficult to confirm since their source is often anecdotal evidence and their relative contributions have not been proven by research. Some also appear contradictory although this in itself is not surprising due to the expected variations in the local corrosion controlling factors on different structures (see Section 2.1).

1. Tidal current: There is some confusion as to the influence of tidal currents. The speed of tidal currents does not appear to affect the incidence or intensity of ALWC (Hodgson, 1998). However, the survey carried out as part of this work (see Appendix 5) indicated that high currents may reduce the risk of ALWC. Strong currents are usually linked to high tidal ranges that can have high silt content, although this is not always the case.
2. Tidal range: The problem has been experienced over sites with variable tidal ranges (Moulin, 1998) although there is statistical evidence that a high tidal range may reduce the risk of the problem occurring (Gubner, 1998). The width of band in which ALWC occurs may also be dependent on the total tidal range.
3. Operations: There is some statistical evidence that high rates of traffic and activity within ports may increase the risk of ALWC occurring (Gubner, 1998).
4. Pollution: A common question posed is whether the incidence and intensity of ALWC is greater in polluted seawater and whether the nature of the pollution (ie from sewage, agricultural, industrial wastes, wind-blown cargo dusts, etc and/or biological products associated with oxygen starved seawater) is a significant factor? For example, pollution arising from the fishing industry is frequently cited as a causative factor (eg Arup, 1963). However, the problem has also been reported in clean marinas and, in one case, its incidence was reportedly reduced to practically zero in an area of a sewage outfall (Hodgson, 1998). While the survey carried out as part of this work (see Appendix 5) indicated that high levels of water contamination can reduce the risk of ALWC occurrence, the nature and type of contamination was not sufficiently well defined to allow detailed interpretation. It is possible that some effluents could act as biocides and others as nutrients, to the relevant bacteria.
5. Sunlight: The influence of sunlight may be a factor since the problem appears to be reduced in shaded areas, or at least its development may be retarded. An example is a quay wall which suffered ALWC but a section of the wall shielded from light survived a further five years before the problem became critical (Shelley, 1998). UV radiation, which can have both lethal and non-lethal effects on bacteria and will penetrate deeper into clearer waters, may be a factor in determining the width of band in which ALWC occurs. The survey supporting this work (see Appendix 5) indicated that only a small percentage of respondees considered there was a link between sunlight exposure and incidence of ALWC and no correlation was found between orientation of the structures and ALWC occurrence.
6. Age of structures: There is evidence of a wide variation in the incidence of ALWC in ports where berths are constructed of steels of different ages. Typically older steel structures have shown fewer incidences of ALWC than newer structures (NCE, October 1998). This may be due to the increased barrier protection afforded by pre-existing deposits present on old steel surfaces formed in the absence of bacterial activity.
7. Steel grade: There is no evidence in the literature connecting the incidence of ALWC with steel grade or manufacturing route although some have linked it with

the change to a "cleaner" grade of steel with fewer impurities in the 1970s (NCE, March 1998). The survey (see Appendix 5) also indicated the type of steel did not have any influence on the incidence of ALWC.

2.3.3 Mechanism

ALWC is associated with bacterial activity (ie with the presence of a consortium of bacteria including SRB and aerobic bacteria) and is a form of MIC as demonstrated by:

- patches of rapid localised (pitting type) attack, characteristic of MIC, seen in other marine applications
- iron sulfide rich corrosion products indicative of SRB activity
- shiny, heavily pitted metal substrate beneath ALWC patches indicative of active corrosion
- the presence of "green rust 2" in the surface deposits previously linked in other studies to corrosion by SRB (Genin, 1993)
- the occasional presence of striations and tunnelling in the steel surface at hole locations, reported as being observed in other marine application involving MIC (Campbell, 1990)
- positive tests confirming the presence of SRB and other bacterial species within the surface deposits in the low water zone
- the occasional presence of tubercles/pustules on the steel surface within the low water zone, previously seen in other marine applications involving MIC (Vrelust, 1992).

Japanese research (OCDI, 1998) appears to discount the role of MIC in low water corrosion preferring to explain such localised corrosion in this zone by macro-electrochemical effects (ie an oxygen concentration cell) with no direct reference to the nature of the surface deposits or their implication on corrosion. In contrast, European research (Johnson, 1997) has placed great importance on both the chemical and biochemical nature of surface deposits found in this zone as an indicator to identifying the cause and concluded "that no other tenable explanation of low water corrosion (ie other than MIC) could be offered, in particular, that macro-electrochemical effects can no longer be considered to offer a viable mechanism".

In the past, it has been commonly assumed that MIC on steel sheet piling was restricted to the mud zone, occurring only at those limited number of sites with contaminated muds. References to isolated examples showing local, high rates of metal wastage (with SRB involvement) in this zone date back to as early as 1948 (Ross, 1948). However, the findings of the present literature survey suggest that MIC is also active and perhaps widespread in the low water zone, although it may take some time to develop, being dependant on water temperature (Melchers, 2003) amongst other things.

For significant MIC to occur the following conditions are necessary:

- the presence of an initial infection
- a supportive growth medium
- favourable environmental conditions for growth
- stagnant zones
- the presence of a consortium of bacteria including SRB and aerobic bacteria.

If MIC in the low water zone is relatively recent on UK structures then it must have a recent cause. Potential causes include:

- harbour muds; providing a source of bacterial infection, for example as a result of dredging activities
- local pollution; providing certain nutrients for bacterial growth, for example, from sewage treatment plants and factories discharging high levels of dissolved organic waste eg fish processing plants, meat packing factories, sugar refineries and so on. Illegal and, therefore, unclaimed discharges from ships would be another potential source of contamination
- ballasting operations; providing a new source of microbial infection or nutrient influx. It is estimated that ships transfer about 10 billion tons of ballast water every year and that more than 3000 species of microbes, bacteria, viruses, animals and plants are carried at any given time (Gupta, 2001)
- "green biocides"; the move away from the use of organo-metallic antifoulant marine coatings on steel surfaces which, while resulting in less toxic heavy metals in coastal waters, may encourage greater bacterial activity in harbours
- global warming; higher water temperatures can lead to rapid development of anaerobic conditions under surface deposits and the development of an SRB population.

If MIC in the low water zone on UK structures has always been evident, then the characteristic black sulfide rich deposits may have simply been mistaken in the past for the black, iron oxide (ie Magnetite) layer formed under "traditional" oxygen diffusion controlled corrosion conditions. The rapid chemical changes that occur to iron sulfides with exposure to air may also have made them more difficult to identify.

Although isolated references to iron sulfide deposits being formed on steel surfaces in seawater date back to the early 20th century, it is only relatively recently that iron sulfide rich layers have been reported on steel surfaces in the low water zone. Several studies on sheet piling have identified both iron sulfide rich layers and aerobic and anaerobic bacteria within surface deposits removed from both inpans and outpans at both ALWC affected and, reportedly, unaffected sites (see Appendix 3). The likelihood is that the iron sulfide forms as a continuous layer on steel surfaces across the whole of the low water zone (rather than forming in discrete patches) and, at some locations, this layer may extend into the continuously immersed zone down to mud level.

The literature also suggests that the existence of such iron rich sulfide deposits is not necessarily associated with high corrosion rates or indeed pitting type attack on immersed steel surfaces (Bultman, 1977). Although the presence of such deposits is one of the characteristics of ALWC, there would appear to be something different happening at ALWC patches compared with the surrounding surfaces in the low water zone. This difference is responsible for the significantly higher rates of metal loss at ALWC patches compared with the more general uniform attack seen at surrounding areas. Several theories have been proposed to explain this preferential attack, including:

- disturbance of the surface deposits in the low water zone exposing the iron sulfide layer to air resulting in oxidation and the production of sulfur (Copenhagen, 1954) which is known to be highly corrosive to steel under certain conditions. Such a mechanism would have serious implications on the normal inspection procedures where surface deposits are routinely removed in this zone for visual examination of the substrate steel and/or for taking thickness measurements

- differences in the sulfate reduction process, the sulfide deposition and the consequent in-situ chemical conditions in the critical corrosion locations, possibly because of differences in the bacterial species (Genin, 1993)
- a significant difference in the nature of the iron sulfide formed at the ALWC patches compared with the surrounding areas in the low water zone due to variations in ferrous ion concentration at such locations as a result of local stress differences, which adversely affects the adhesion and protection characteristics of the surface deposits formed (Breakell, 1998)
- local production of acids by SOB in addition to sulfides (Gehrke, 2003) or the biochemical breakdown of dead organisms
- formation of differential aeration cells; either by local shielding of surfaces by marine biofouling producing localised anaerobic conditions and tubercle formation (Vrelust, 1992) or dissolved oxygen differences between the tidal and low water zones (Tsuruta, 1973) particularly if the low water zone surface deposits are regularly disturbed (Marsh, 2003).

The literature survey findings also suggest that ALWC may not be a new problem, although its geographic distribution or development is unclear. Whether the problem is more recent in the UK, as the literature would suggest, is uncertain. If the problem has been longstanding, then it is difficult to explain why the distinct visual characteristics of ALWC (ie "orange" patches) and their significance (ie rapid metal thinning, holing, loss of backfill) had not been detected and reported much earlier on, at least on some structures. It may be that the problem was simply never looked for in the past. There has been (and still is) a general lack of knowledge about the condition of sheet piling in the low water and submerged zones and previous inspections in these zones are known to often have been "cursory" (Holleywood, 1978). There was also a general belief in the past that marine biofouling in the low water zone provided some degree of protection against corrosion. It has not been appreciated, until more recently, that such marine biofouling can, over time, also provide a suitable environment for colonisation by micro-organisms capable of affecting the corrosion process, in particular allowing anaerobic bacteria (eg SRB) to flourish beneath. Once this biofouling layer becomes sufficiently thick and bio-active, anaerobic conditions are established at the metal surface and the corrosion process becomes bacterially controlled (Gehrke, 2003; Bultman, 1997). If such anaerobic conditions are not maintained, then oxygen diffusion controlled corrosion of steel would be expected to take place and iron sulfide layers would not be formed.

Although uncertainty still remains over the precise mechanism for ALWC, it is probable that it is a multi-staged process involving an induction (or incubation) period (Melchers, 1997 & 2003; Gehrke, 2003), possibly lasting years on some structures, with subsequent stages producing attack at different rates which may be linear, exponential or stop-start, depending on circumstances. The rate-controlling step in the mechanism, responsible for driving the most rapid rate of metal loss at ALWC patches, may be a result of ongoing bacterial corrosion, which is known to be capable of producing rapid local metal wastage on steel in other marine applications. A contribution from a macro-cell effect at some stage, arising from the surrounding significant area adjacent to the ALWC patches within the low water zone itself and/or between the tidal and low water zones, can also not be discounted.

2.4 Corrosion rate data

Before summarising the literature findings on corrosion rate data for steel in marine environments, it is necessary to comment briefly on corrosion measurements. This is because it is not always clear how the values reported in the literature are derived or indeed whether the value quoted is referring to general (uniform) or localised corrosion. As a "rule of thumb" in marine corrosion, localised corrosion rates are typically quoted as 1.5 to 3 times those given for the general (uniform) rates.

Corrosion measurements are often reported in terms of general corrosion, where the loss of metal is assumed to be evenly distributed over the entire metal surface. The values presented are normally obtained onsite by ultrasonic thickness measurements (see Appendix 7) or from metal weight loss determinations on de-scaled test coupons. They assume general corrosion is occurring at a linear rate over a certain exposure period. Care should be taken when using such corrosion data to predict when holing will likely occur. This is particularly so when it has been derived:

- using short term exposure periods where any protective surface deposit build up has not had time to reach a "steady state" condition
- from small weight loss test panels fixed to, but electrically isolated from, structures since they do not take into account the effects of any concentration cells that may be in operation on the structure itself.

The first three columns in Table 2.1 provide a summary of typical corrosion rate values for steel per exposed side in the various zones together with a commentary reflecting the accepted "traditional" corrosion understanding of rate controlling factors. Within these zones a range of notional average and upper limit values of corrosion for exposed, unprotected structural steels in temperate climates are given from a variety of sources as a guide for designers to what could be expected. It will be seen that the actual rates given do vary from source to source and reflect differences in exposure conditions at the locations where the measurements were taken, the age of the structure and how the measurements were obtained.

Values reported for ALWC corrosion rate from various referenced sources are presented in the last column of this table associated with the "intertidal low water zone" (ie the highlighted cell). The corrosion rates in the continuously immersed zone exclude any ALWC effects as they have only recently been reported to occur in this zone and no quantitative corrosion data is available in the literature. Care should be taken when interpreting the range of corrosion rates given for ALWC. This is because ALWC is a multistage process involving an induction period, initiating at some unpredictable point in time (Sonck, 2002). The assumption typically used in the literature to estimate ALWC corrosion rates is that the corrosion occurs at a linear rate over the life of the structure. The rates given may, therefore, be underestimates of the actual rate (ie the instantaneous rate) occurring on the structure in real time (see Section A, Appendix 6).

Table 2.1 Typical corrosion rates found in the literature

Environment and commentary reflecting the "traditional" understanding of factors controlling corrosion rate (see Figures 1 and 2.1)	Average/mean corrosion rate (mm/side/year)	Upper limit corrosion rate (mm/side/year)	ALWC corrosion rate (mm/side/year)
Atmospheric zone *– the area between splash zone and top of the structure.* Exposed to a salt laden atmosphere.	0.04[(b)] 0.04[(e)] 0.035[(f)] 0.019[(g)] **(Range: 0.02 to 0.04)**	0.10[(e)] 0.41[(g)] **(Range: 0.10 to 0.41)**	
Splash zone *– the area from mean high water (MHW) up to a height depending on the mean wave height.* This zone is normally where the highest corrosion rates would traditionally be expected to occur although the actual rate will usually depend on exposure of the structure to wind and wave action.	0.08[(b)] 0.15[(c)] 0.08[(e)] 0.075[(f)] 0.42[(g)] 0.3[(h)] 0.1 to 0.2[(i)] 0.4[(j)] **(Range: 0.08 to 0.42)**	0.17[(b)] 0.17[(e)] 0.3[(i)] **(Range: 0.17 to 0.30)**	
Tidal zone *– the area between MLWS and MHW.* A particularly low corrosion area exists in this zone typically about halfway between MHWS and MLWS. An explanation for this is the formation of a differential aeration cell (AISI, 1981). The steel surfaces in this zone are intermittently exposed to highly aerated seawater as the tide flows in compared with the reduced oxygen content of the seawater in the relatively stagnant low water zone. On some structures marine biofouling or floating oil films can coat the steel surface in this tidal range. The presence of such layers has also been considered to provide partial protection within this zone and may also be expected to reduce any differential aeration effect between zones.	0.04[(b)] 0.08[(c)] 0.04[(e)] 0.035[(f)] 0.1[(h)] 0.0 to 0.1[(i)] 0.05[(j)] **(Range: 0.04 to 0.10)**	0.10[(b)] 0.18[(i)] 0.10[(e)] 0.15[(j)] **(Range: 0.10 to 0.18)**	
Intertidal low water zone (ILW) zone *– the area between a point approximately 0.5 m below LAT to MLWS.* A significant metal loss can occur at short distance below MLWS traditionally attributed to the action of a differential aeration cell as described above. The presence of discontinuous marine biofouling in this zone may produce enhanced localised corrosion. Also, repeated local removal of the semi-protective surface deposits in this zone can result in enhanced localised attack.	0.08[(b)] 0.08[(e)] 0.075[(f)] 0.15[(i)] 0.2[(j)] 0.08[(k)] **(Range 0.08 to 0.20)**	0.17[(b)] 0.17[(e)] 0.33[(i)] 0.34[(k)] **(Range 0.17 to 0.34)**	This is the zone where ALWC and associated high rates of corrosion are most commonly seen. 0.5 to 0.8[(e)] 0.3[(h)] >0.3[(i)] 1 to 4[(L)] 0.1 to 1.2[(m)] **(Range 0.3 to 4.0)**

***Continuous immersion zone** – the area from mud line (sea bed) up to the lower limit of the ILW zone.* Traditionally, a low corrosion area attributed to the build up of semi-protective rust layers (scale) and marine biofouling.	0.13[(a)] 0.035[(b)] 0.04[(e)] 0.035[(f)] 0.1[(h)] 0.03[(i)] 0.075[(j)] 0.04[(k)] **(Range 0.04 to 0.13)**	0.13[(b)] 0.13[(e)] 0.2[(h)] >0.1[(i)] 0.15[(j)] 0.14[(k)] **(Range 0.13 to 0.20)**	ALWC with its associated high rates of corrosion have recently been reported in this zone on some structures. No quantitative corrosion data is available in the literature.
***Mud (embedded) zone** – the area of total burial in mud or soil.* This area is normally considered to suffer minimal corrosion unless contaminated with pollutants or high levels of SRB.	0.03[(h)] 0.05[(i)] 0.075[(j)] **(Range 0.03 to 0.08)**	0.015max[(b)] 0.015max[(e)] >0.05[(i)] 0.1[(j)] **(Range 0.02 to 0.10)**	
Backfill: • above the residual water level • below the residual water level.	0.03[(h)] 0.02[(h)] <0.2 [(i)]		
In flowing seawater (in the absence of fouling organisms) • seawater at 11°C • seawater at 21°C • seawater at 25°C.	 0.11[(a)] 0.36[(a)] 0.51[(d)]		

(a) L L Shrier, "Corrosion", Butterworth-Heinemann, Vol. 1 p 2.51

(b) British Steel Corporation Piling Handbook, Feb. 1997

(c) M G Fontana and N D Greene, "Corrosion Engineering", p268

(d) "Corrosion Data Survey", NACE (5th Edition), 1981

(e) BS 6349-1: 2000, Table 25

(f) Corus brochure "Durability and protection of steel piling in temperate climates" 2002

(g) "Handbook of Corrosion Protection For Steel Pile Structures in Marine Environments" American Iron and Steel Institute (AISI), 1981

(h) "Corrosion Protection and Repair Manual for Port and Harbour Steel Structures" The Overseas Coastal Area Development Institute of Japan, 1998

(i) "Inspection, maintenance and repair of maritime structures exposed to damage and material degradation caused by a salt water environment" PIANC Report of Working Group 17, 2004

(j) "A guide to the selection of marine materials" INCO databooks 1973.

(k) M Graff "Accelerated Low Water Corrosion (ALWC) and Microbial Induced Corrosion (MIC): Two Sides of the Same Coin?", ALWC Conference, Denmark 2002

(l) "Maritime Board briefing on ALWC", The Institution of Civil Engineers.

(m) The survey supporting this work, 2004

Summary

- "traditional" (ie oxygen diffusion controlled) marine corrosion occurs on unprotected steel structures, which varies in intensity depending on the position on the structure.
- ALWC is a localised corrosion phenomenon that typically occurs just above LAT. It has a strong microbial component that likely drives the corrosion process. It is associated with unusually high rates of metal wastage, specific visual features and poor adhesion characteristics of the surface deposits formed
- ALWC is now probably the most likely, common and aggressive form of corrosion to be found in the low water zone but it is not the only form of corrosion that can produce localised attack on maritime structures in this zone
- a summary of researched corrosion rate values for both general (uniform) and localised corrosion (including ALWC) is given in Table 2.1.

3 Geographic distribution of ALWC

3.1 Geographic areas at risk

There is general acceptance that the ALWC phenomenon is global, having been reported around the world in all climatic conditions. It can occur on any unprotected steel structure, on inadequately protected structures or where protection has broken down, in contact with saline water such as seawater and brackish water which is subject to tidal influences.

ALWC in the UK was first reported as a significant problem by port authorities and consulting engineers mainly around the coasts of southern England and the North Sea. Of these, approximately a dozen UK sites in the 1990s are known to have needed major repairs as a result of this problem, for example, a £2 million renovation project was undertaken in Southampton in 1996 (Parker, 1998). The problem has also been reported at a number of sites in Scotland (Carmichael, 1998) and Northern Ireland where remedial works have become necessary much earlier than expected.

Locations in European countries outside the UK where the problem has been reported include Scandinavia, Belgium, Germany (Hamburg and harbours located both in the North Sea and the Baltic Sea), Holland and France. Cases of ALWC have also been confirmed in the USA, Canada, Zanzibar, South Africa, Australia, Cyprus, UAE, Japan and in the Southern Caribbean.

3.2 UK surveys

The extent and spread of ALWC in the UK is a source of ongoing debate, although several surveys have been undertaken in the past in an attempt to clarify the issue.

A survey was performed covering five Western European countries, including the UK, as part of a study by the European Coal and Steel Communities (ECSC). It was reported that three to four per cent of steel sheet piled structures (Moulin, 1995) and at least 13 per cent of the ports and harbours were affected by ALWC (Johnson, 1997). In conclusion, it was assumed that ALWC was not present at sites from where no questionnaire was returned.

A further survey of UK ports and municipal structures was initiated in 2001 by the Permanent International Association of Navigation Congresses (PIANC) as part of a census organised by the Maritime Board of The Institution of Civil Engineers (Shelley, 2002). Of the three hundred questionnaires sent out, seventy four replies were obtained of which thirty four cases of ALWC were reported.

A survey was also initiated as part of this study and the questionnaire is reproduced in Appendix 5. The objectives were:

- to obtain more detailed information from a greater number of port authorities than was gathered during the ICE survey identified above
- to gain a better understanding of ALWC, its geographical spread and identify potential factors that may be contributing to its occurrence.

Of the seventy questionnaires received as a result of this survey some forty reported that ALWC had been identified and a further fourteen reported that corrosion in the critical zone had been observed but there was uncertainty as to its cause. Sixteen reported that no significant corrosion was present. Additional information obtained from this survey, as was available at the time of publication, has been incorporated into this document and a preliminary analysis of the data obtained is presented in Appendix 6.

The areas of UK coastline reported to be affected by ALWC are shown in Figure 3.1 below. This map of reported ALWC sites in the UK is based on the combined findings of previous survey responses, the survey initiated as part of this study and information from both published and unpublished sources. Anecdotal evidence is available to suggest that ALWC is also present at other areas not marked. Note that locations have not been indicated exactly and large diameter location points have been used to protect the site identity.

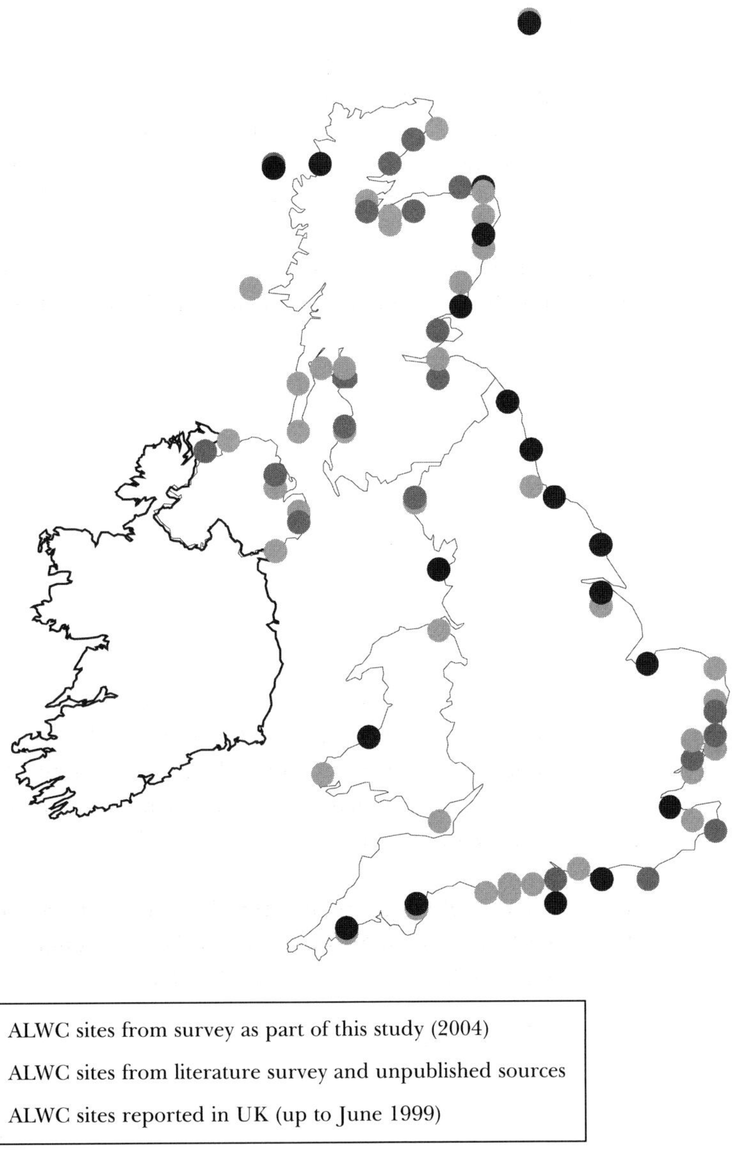

Figure 3.1 **Reported ALWC sites in the UK**

Overall, accurate figures to identify the scale of the problem and its progress have proved difficult to establish with any certainty largely due to the limited responses to the questionnaire surveys, resulting from factors such as:

- a lack of understanding of the ALWC problem and its identification
- commercial sensitivity
- structures not yet having been adequately inspected for ALWC in the critical zone
- structures constructed in materials other than steel (eg granite and masonry) so not relevant.

Those that have responded to questionnaires are typically from ports and harbours where regular low water inspections are undertaken, facilitated by the sheltered nature of the waters at such locations. This, together with the general acceptance of circumstantial evidence and hearsay, suggests that the problem is more widespread than reported here.

Summary

- there is general acceptance that ALWC is now a global problem
- ALWC can potentially occur on any insufficiently protected steel structure in contact with saline brackish water which is subject to tidal influences
- the questionnaire-based surveys completed to date suggest that not all sites are affected by ALWC
- there is no location where operators or designers can rule out the possibility of ALWC occurring.

4 Asset management for ALWC

This chapter presents general principles of asset management in a maritime environment with particular emphasis on ALWC. Asset management is intended to preserve a structure in a safe and operable condition with optimised spending on maintenance, and is equally relevant to owners of large and small assets. Each management method should be tailored to the specific requirements of each site, requiring the input from owners, engineers and designers of the asset. Differences in the approach to asset management between large and small asset owners, and health and safety issues are addressed in this chapter.

Deterioration due to ALWC on a structure leads to loss of performance including reduced load capacity and, in the case of retaining structures, loss of fill material due to holing of the structure. Ultimately, the integrity of the structure and potentially that of adjacent structures is at stake. To obtain an estimation of when this point will be reached (when timely action is required) and to determine appropriate countermeasures, it is necessary to manage the sometimes large amount of data that is collected in surveys. Therefore, an effective data management system is an important tool in asset management. These systems can be linked to predictive modelling tools including asset costing and assessment techniques, which will enable owners to plan and budget future maintenance and inspection and the engineer to design the optimum solution for each specific problem. However, much simpler methods can be as effective in small ports. The survey carried out as part of this work (see Appendix 6) has shown that only 12 per cent of ports and harbours currently have an asset management system (including planned inspection and maintenance) in place. Many responses to the questionnaire began to keep inspection records from the mid 1990s.

In principal, there are two basic types of maintenance strategy: proactive and reactive. A structure that is subject to proactive maintenance undergoes regular renovation in order to avoid future damage. A reactive strategy, in contrast, limits surveys and repair to directly observable damage.

4.1 Consequences of loss of performance

ALWC influences the performance of steel maritime structures such as sheet piled walls and may compromise operational safety leading to disruption of facility operations and subsequent loss of revenue. Asset management is used to predict and help to allocate sufficient maintenance resources to allow efficient and safe operation of structures within budgetary constraints. When asset management is implemented properly, it will be a tool to forecast not only actual maintenance cost but also losses in revenue due to downtime and matters affecting safety.

The safety consequences for a structure reaching its ultimate limit state by ALWC can be significant and, in this case, potentially render the structure unusable (see Section 1.4). While holing of retaining steel piling may not have structural implications initially, the washout of fill material caused by the tidal flow in and out of the holes may result in lack of support and ultimately to failure of the structure, sometimes without warning. In other instances, the perforation might weaken the structure causing flexural failure (bending) of the steel sheet piles and subsequent collapse of the structure.

These problems can only be identified and hence avoided by planned and scheduled inspection (prescriptive inspection), monitoring and maintenance regimes. These regimes should be planned for and applied to:

- new build structures when the whole life cycle cost of the structure is calculated at the design stage, thus providing a figure of the overall cost of the structure during its lifetime
- existing structures when an assessment is made to determine residual life of the structure.

This rigid approach to inspection and monitoring is appropriate for use on, for example, a small port. It can be improved/superseded by more modern performance based inspection regimes, where the frequency of inspection depends on the actual state of the structure and the outcome of predictive asset management models. The latter approach is the preferred option for owners say of large ports due to cost and time savings. A diagram depicting a generic asset management method applicable to all port sizes is shown in Figure 4.1.

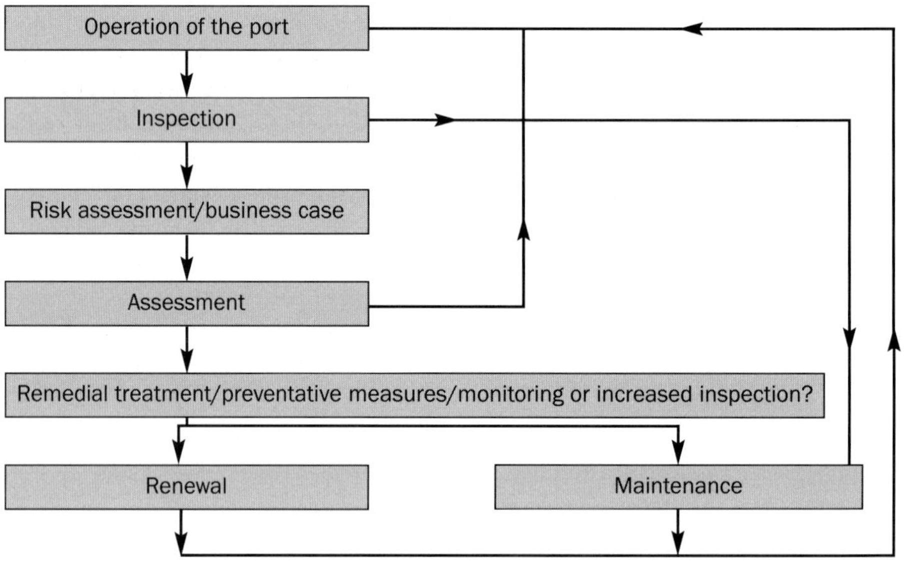

Figure 4.1 **The asset management cycle**

Efficient and effective management of maritime assets is essential if performance requirements are to be met. These requirements include:

1. Safety and reliability.
2. Operational efficiency.
3. Providing assurance and confidence in the operation of the asset for all stakeholders.
4. Satisfying customer and employee expectations and perceptions.
5. Satisfying statutory and regulatory obligations.
6. Value for money and business improvement.
7. Minimising environmental impact.

Additionally, benefits of an effective asset management system include:

1. Converting owner policy and objectives into appropriate actions.

2. Assisting with the prioritisation of expenditure.
3. Providing comparative analysis between assets.
4. Supporting submissions to financial sponsors for funding maintenance works.
5. Providing information regarding the serviceability and rate of improvement or deterioration of the asset.
6. Assisting with the quantification and mitigation of risks.
7. Identifying immediate and future investment requirements.

4.2 Legal implications under Health and Safety legislation

Owners and operators of maritime structures (clients) have legal obligations to maintain the asset in a safe condition to protect employees, the public and third parties from unacceptable risks. These obligations arise from statute and licence conditions and also from common law and ethical considerations (duty of care). Combined, they have a significant impact on asset management policy and operational procedures.

In the UK, the principal statute governing safety is the Health and Safety at Work Act (1974) being enforced by the Health and Safety Executive (HSE). The CDM Regulations 1994 impose a duty on clients, designers and contractors to have adequate regard to the possible effects of design on the health and safety of those who will carry out the construction, maintenance or demolition of the works as well as others who might by affected by their activities. It is a requirement of the CDM Regulations that designers must consider, during the development of their designs, the hazards and risks which may arise to those constructing, cleaning and maintaining the structure. It is also a requirement that at the end of construction a health and safety file is produced. This file is an essential reference document for any subsequent work on the structure including cleaning and maintenance. Clients should also ensure that designers and contractors are competent to carry out the tasks for which they are employed. Having a well managed asset management system will potentially provide some defence if an incident did occur. It is likely to become increasingly unacceptable not to do so and not to have a good appreciation of the condition of your structures.

4.3 Economic considerations

In terms of asset management, ALWC is part of the ageing process that has not until recently been recognised. It can result in unforeseen expenditure and, if not properly addressed, it will make the structure economically unsustainable.

In general, asset risk management strategies are set up to protect the value of the structure at an economically sustainable level (Figure 4.2). The management used to implement these strategies can be as simple as the evaluation of inspection records, but they might also consist of more sophisticated systems such as databases incorporating predictive models. Maintaining the structure at an economically sustainable level is achieved by optimising the costs to the user and owner of the capital and maintenance costs.

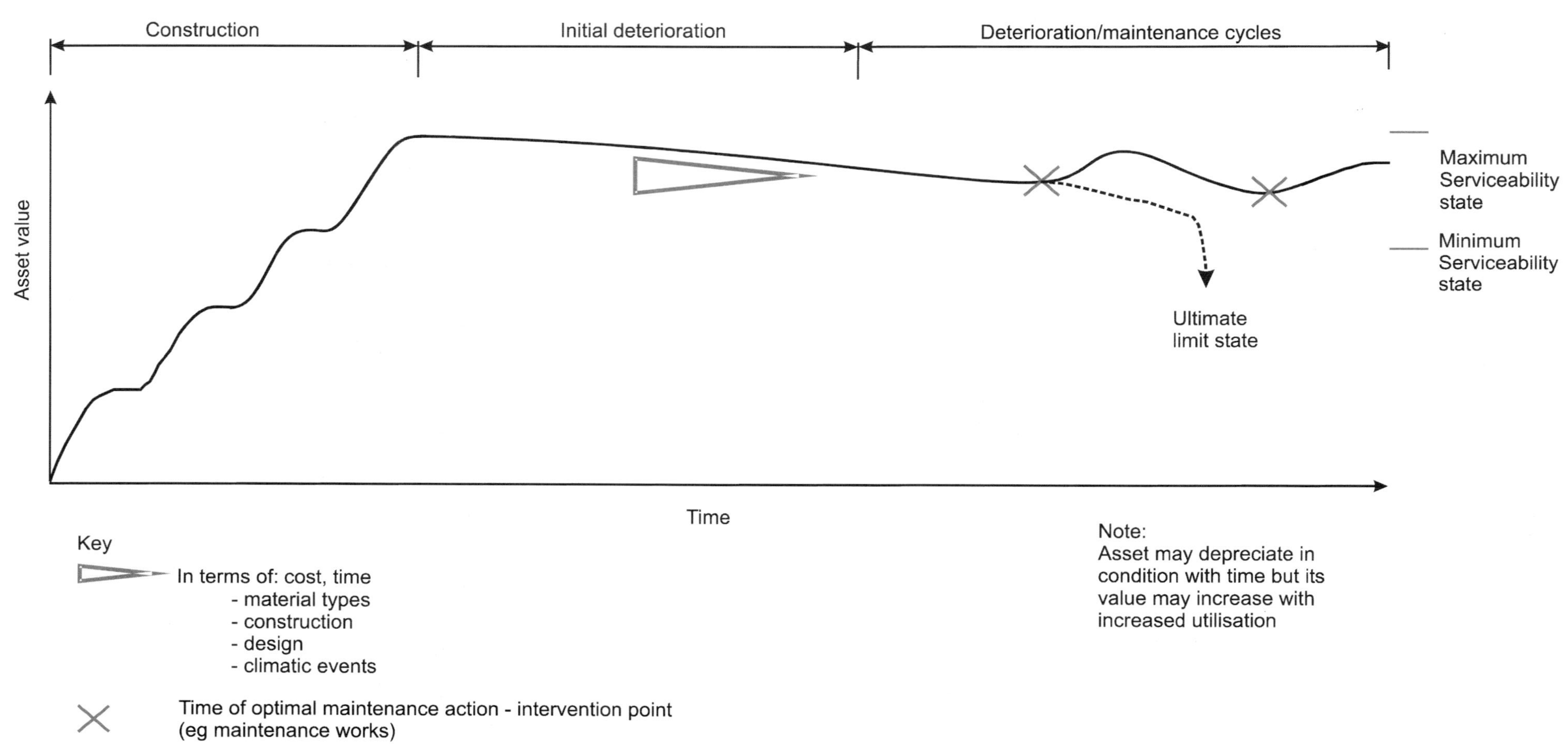

Figure 4.2 **Fixed asset value model**

The expenditure on maintenance, the cost to users and the benefit gained is monitored using key performance indicators, to ensure the asset remains within serviceability limits both physical and financial. Key performance indicators are not prescribed but have to be defined for each asset according to the needs of owners, operators and users. For example, costs per year, disruption to facility operation, frequency of painting and so on.

4.4 Whole-life asset cost (cost effective long-term management strategies)

Whole life costing can be a useful exercise at the onset of a project to determine the optimum solution in terms of cost effectiveness. It is also useful when comparing the capital cost of renewal schemes against longer-term maintenance costs. Whole life costing of a structure consists of the following elements:

- client decision layout based on concept designs
- detailed design and preparation of construction data
- construction of structure and commissioning
- monitoring, inspection and maintenance works
- decommissioning – demolition.

The following parameters are required to perform the financial analysis in whole life costing:

- capital costs (including design costs, construction statutory procedures, etc)
- the life cycle of specific costs, ie when do these costs re-occur
- future maintenance costs
- residual life
- estimated residual life without remedial work
- discount rate.

The output of whole life costing is variable as results depend on assumptions made in the model, in particular the discount rate. Generally, a higher discount rate stimulates low cost construction and early replacement. Discount rates will vary from country to country or even from operator to operator. The choice of the discount rate results in different design qualities of the infrastructure and, therefore, this rate needs to be determined from the start (decision layout).

One major criticism of whole life costing is that it attempts to predict future events but with time the forecasts predicted by the model will become less accurate. Since ALWC is a relatively new phenomenon forecasts made by previous models will have to be reconsidered and adjusted regularly and as necessary. An owner, when faced with ALWC where maintenance is too costly, will need to consider the renewal cost against what is almost certainly unacceptable loss of trade.

4.5 Residual capacity and service life assessment

Current design and assessment codes of practice require that consideration is given to ultimate and serviceability limit states. These terms are defined as follows:

Service life and serviceability limit state

The serviceability limit state is defined as the state of deterioration of a structure such that its use is affected, its durability is impaired or its maintenance requirements are substantially increased. A typical example with ALWC is excessive thinning and holing, giving rise to additional maintenance effort to allow for future operation of the structure. In most instances, defects that impact on serviceability will ultimately generate a limit state problem if there is no maintenance intervention or the design life is exceeded. It is often inexpensive and also reduces the risks (health and safety, and other liabilities) when such problems are resolved before the ultimate limit state has been reached.

Ultimate limit state

The ultimate limit state is defined as the state of collapse, instability or forms of failure that may endanger property or people or cause major economic loss. At the end of the design life of a structure, major remedial works or renewal of the structure may be required. Again, ALWC may have progressed so far that holes appear in the structure and washout of backfill has commenced. At this stage urgent, expensive, remedial action is required. However, it might be possible by assigning a long design life (ie by introducing extremely high construction standards) to eliminate all but routine maintenance.

It is difficult to foresee future conditions, such as ALWC, which have a significant impact on the service life and, as a result, there is always a certain degree of uncertainty. In asset management systems, degradation mechanisms are usually addressed by assigning deterioration rates that will allow a forecast on the likely mechanical performance and, hence, section thickness required.

4.6 Maintenance strategy

The choice of maintenance strategy (ie preventative inspections on an "as needs" basis) will depend on factors such as the type and size of structure, the available budget and on other resources available to the maintainer. Several potential approaches have been introduced in this chapter but ultimately the client will need to make the appropriate choice of maintenance strategy based on their circumstances. The basis of each maintenance strategy is the data upon which it is built. Appropriate data management is, therefore, paramount to a good maintenance strategy.

4.6.1 Data management

The need for data management will be different for each facility and will depend mainly on the size of the facility and the chosen maintenance strategy.

The collection of data is only useful if it is then processed and used to provide an indication of the condition of the structure. Knowing this and the associated operation and maintenance cost is the key to monitoring and maintaining the asset between serviceability limits (Figure 4.2). When the inspection data is used properly, adjustments and refinements can be made on future needs and expenditure of the asset.

However, very often inspections are carried out and the records are not used any further. It can be very tedious (and expensive) to filter out unwanted information from critical information and it is, therefore, considered good practice to limit the amount of data generated by each inspection to a minimum. This can be achieved by focusing on critical elements or on recording relevant data (related to some key performance indicators). However, in this instance, the inspector needs to be sufficiently qualified to make a decision as to what data is relevant on site. Incorrect or inappropriate data can be misleading.

A frequent inspection and maintenance regime generates a significant volume of data, which will only be of benefit if is readily accessible and provides useful information to the engineer, owner and operator, thus enabling them to direct appropriate maintenance action. Information technology has facilitated the implementation of effective data management and analysis tools. These are very often tailored to and developed for the specific needs of owners of the infrastructure.

4.7 Risk assessment methods

Risk assessment is needed for a number of reasons (see Sections 2.4 and 6.3):

- to satisfy statutory safety obligations
- to understand and manage the risk of impact on the facility operation including financial risk
- to control financial implications of risk to operation.

The Management of Health and Safety at Work Regulations (1999) impose a statutory requirement on employers to undertake risk assessments. This is to ensure that safety is maintained at all times and not degraded, and that risks are averted or reduced according to a principle called ALARP (as low as reasonably practical). This principle assures that safety improvements are pursued so that the cost of averting the risk is proportionate to the risk.

Risk assessment is a tool that can be applied by maintenance managers and clients to ensure that safety, operational and financial objectives are met within the business framework and that funds are justified and allocated appropriately.

There are two basic levels of risk assessment, namely strategic and tactical. At strategic level the objectives of the assessment are identified followed by the identification of hazards (or events), which could prevent the achievement of the objective. At tactical level each perceived risk is entered into a risk register and the cost of mitigation or reduction to ALARP level is determined for each risk. This register allows the user to make a decision on the optimum level of spending while reducing the risks to an acceptable level. It is also good practice to have certain contingency procedures in place should a certain event occur. These procedures (and the risk assessment) should be reviewed at frequent intervals (eg every three to five years) or on "as needs" basis, when significant changes have occurred.

Once the risk mitigation measures have been selected and implemented, the structure should be monitored within the overall inspection/monitoring programme. Risk assessments are often carried out as desktop studies or as workshop activity in which the probability of failure and consequences are determined through the combined experiences of the assessment team. Following this, inspections of critical elements of the structure can be conducted, which serves the purpose of determining the condition of the asset and give a benchmark against which deterioration (or improvements) can be evaluated. Inspections have to be planned carefully as data will be useful in risk assessment only if it can help to identify either the probability or the consequences of a hazard or both. They also need to be focused (ie to relevant parts of the structure, obtaining specific types of data collected at appropriate intervals) so that the amount of data collected remains manageable.

The approach to risk assessment depends on the size of the inventory and can vary substantially for small and large ports. However, it is possible to define a generic risk assessment strategy for structures with ALWC (Figure 4.3).

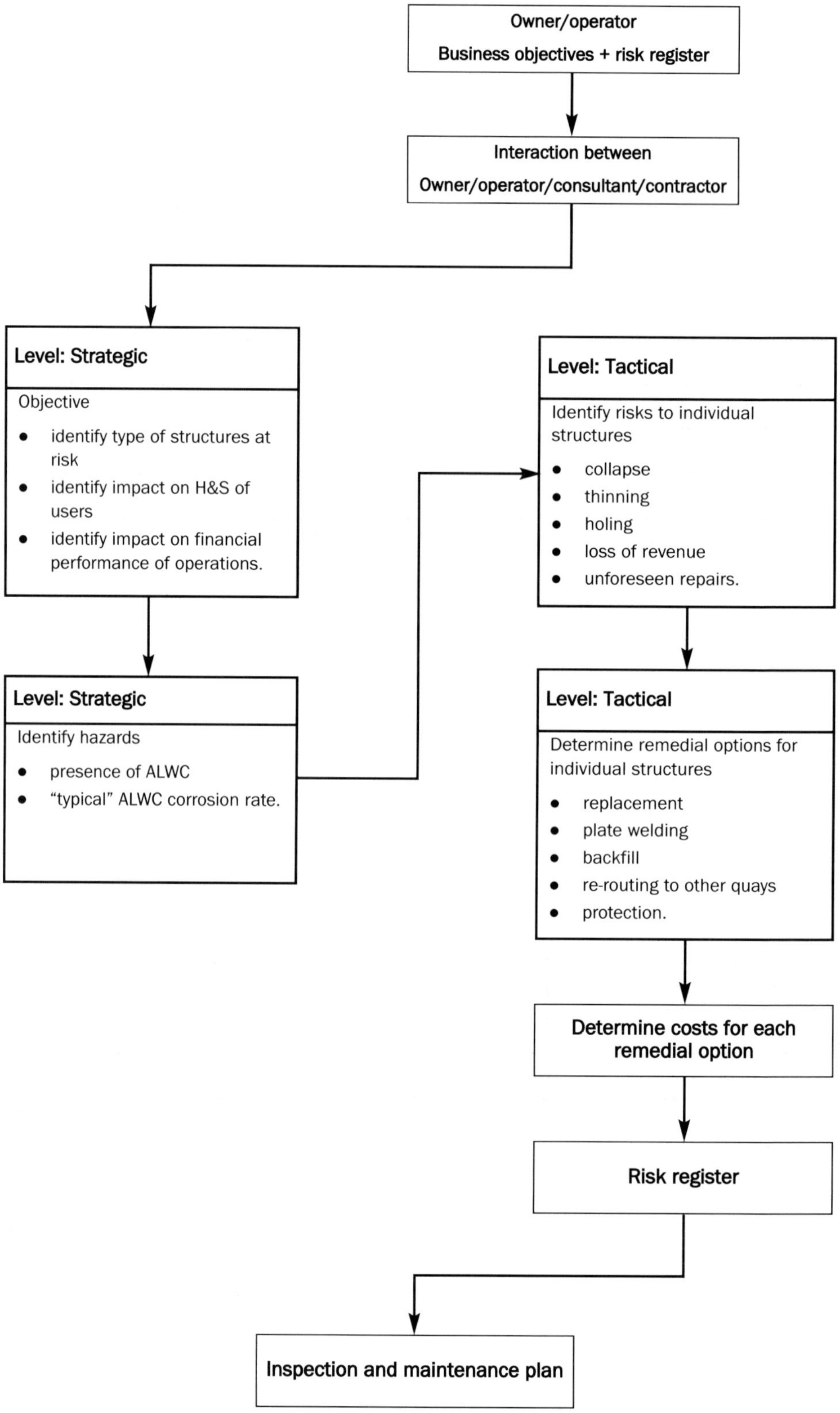

Figure 4.3 Risk assessment procedure for ALWC

Further information on general risk assessment can be obtained from CIRIA SP125 "*Control of risk*" (CIRIA, 1996).

4.7.1 Risk assessment methods – small asset owner

Initially, critical elements require identification (desk study or workshop), typically in collaboration with an expert (eg qualified engineer), with possibly mechanisms and damage functions included and an intervention level established below which action is required. This could be taking steps to watch out for first signs of ALWC (at anytime) and, once it is discovered, commissioning a thorough survey (eg by boat and/or using divers).

A simple, but formalised method of risk assessment consists of assembling a matrix with different parameters, such as remaining section thickness, to which a value is assigned according to the state of the asset. The need for maintenance action is then determined based on the accumulative result of this risk assessment. An example of risk evaluation matrix for simple port structure is presented below in Figure 4.4.

Evaluation matrix		
Condition state		**Action**
1	As new.	
2	Slightly degraded paint system, no corrosion/other signs of usage.	
3	First signs of corrosion/degradation/fractured plates.	
4	Widespread corrosion 30-50% loss of section.	Initiate design check
5	70% loss of section.	Action needed
6	80% loss of section and/or signs of ALWC.	
7	Holing, deformation.	
8	Collapse.	

Asset description	**Condition state**					
	1980	1985	1990	1995	1996*	2000
West wall	1	2	6	6	1	1
East wall	2	3	6	6	1	1
Fittings	2	3	4	4	1	2
Quay surface	2	3	3	8	1	2
* major refurbishment and replacement of quay walls and resurfacing						

Figure 4.4 **Example of condition assessment matrix for simple port structure**

4.7.2 Risk assessment methods – large asset owner

The owners of large inventories need to put in place a formalised, structured and economical system for risk assessment. Once critical elements have been identified, modelling of material and structural degradation of these elements will determine the frequency of inspections and the need for future action. A simple model can be constructed using a spreadsheet programme (with the appropriate corrosion rates) to determine the impact on structural safety and the expected beneficial effects of applying corrosion protection measures.

Deterioration models can be developed to calculate the probability of a given member being in a certain condition state after a particular age. Degradation processes of steel structures in a marine environment are mostly corrosion related. The standards and guidance documents give deterministic corrosion rates for certain exposure conditions in mm/year/surface of exposed steels. In the case of ALWC, however, this approach is over simplistic, as corrosion does not occur at a linear rate over the life of a structure and published values to date are probably underestimates of actual rates. Therefore, in the absence of actual measurements on a structure, the mid to higher end of the range of corrosion rate values given for ALWC in Table 2.1 corresponding to the ILW zone should initially be used to predict the loss of section in a corrosion model when ALWC is observed. It is important to note that using this approach is more likely to result in a 100 per cent overestimate of section loss much more often than a 50 per cent underestimate.

There are a variety of risk assessment methods available. A suitable method for the analysis of maritime structures is given in the "Failure Mode and Effects Analysis" (BS 57605 1991). This is a procedure by which each potential failure mode of a system is analysed to determine the effects on the system and classify each potential failure mode according to its severity. For example, in the case of ALWC on a quay structure, this would mean that ALWC causes holing (after a period of time to be determined from a corrosion model) which in turn causes washout of the backfill and collapse of the structure. Additionally, the structural integrity might be impaired if the holing occurs at regular space intervals. This should be determined on an individual basis and will be dependent on the findings of a detailed survey and subsequent structural analysis (but only if the corrosion pattern suggests that this is needed).

Summary

- where ALWC is identified on a structure, then it ceases to be a risk but becomes a problem that can be managed
- asset management enables objective evidence of potential asset and operational losses to be forecast and presented to management when allocating budgets for remedial work
- implementation of an asset management system for all asset owners is essential in dealing with ALWC in order to increase economical efficiency and optimise expenditure
- asset management follows a continual process of inspection, assessment and remedial treatment (see Chapters 5 and 6), leading to a continual awareness and understanding of (and improvement in) infrastructure condition
- ongoing condition appraisal documentation for each structure should be maintained and analysed to predict when maintenance and repair will be needed as part of an asset management programme
- each management method needs to be customised taking into consideration the specific requirements for each site. This requires input from the owners, their engineers and designers of the asset
- irrespective of size, all owners have a duty to maintain their assets in a safe condition. However, differences in the approach to asset management are suggested in this chapter relative to the size and resources of the asset owner.

5 Condition appraisal

In general, the earlier ALWC is detected and addressed the less extensive, and expensive, the remedial solution. The condition appraisal provides the information needed to:

- identify if there is an ALWC problem, or the potential for one, on a particular structure
- evaluate the extent and nature of this problem so that the structure's remaining serviceable life can be predicted, based on the risk or impact that it might have for one or more components of that structure
- identify and prioritise any remedial treatment that may be required for the structure in order to limit or avoid any further deterioration.

This chapter considers the requirement, and then presents a methodology, for the condition appraisal of the structure. Following a consideration of the personnel who should undertake the work, each stage of the appraisal is described:

- desktop study
- planning
- initial inspection
- preliminary analysis
- detailed survey
- detailed analysis.

Ongoing monitoring of the structure is then considered. A detailed description of relevant surveying and monitoring techniques, including means of access, is presented in Appendix 7. Also see Section 1.5 for Health and Safety requirements.

5.1 Requirement for condition appraisal

The condition appraisal is important as it allows:

- the identification and quantification of problem areas or potential problems
- maintenance and protection to be prioritised
- funds to be properly targeted
- business and safety risks to be quantified and addressed.

Condition appraisal of both new and existing structures is part of the asset management cycle (see Chapter 4). While such an approach is clearly desirable, the survey supporting this report (see Appendices 5 and 6) suggests that many port owners do not currently implement any regular, frequent inspection regime, conducting inspections only as part of remedial works. This reactive, rather than proactive, approach is aggravated by the absence of a specific legal requirement for inspection of maritime structures in the UK other than through the Health and Safety legislation in force.

5.2 Appraisal methodology

5.2.1 Methodology

Inspecting for the presence of ALWC need not be unduly expensive, difficult or time consuming and, if discovered in its early to intermediate stage, does not demand immediate expenditure. If the problem is detected early enough, monitoring of the situation by regular re-surveying, while planning an appropriate protection method or future repair/maintenance schedule, would be an adequate response.

Inspections and surveys are of three types:

- initial and monitoring inspections; usually completed at regular intervals in accordance with a predetermined programme. The first inspection may often be considered to be a "baseline" inspection
- detailed surveys; performed when initial or periodic monitoring inspections indicate a requirement for a more detailed assessment. The frequency of these surveys may be expected to increase with the age of the structure
- work-related inspections; detailed surveys specifically performed before, during or after remedial work.

A suggested condition methodology regime for a typical port structure is given in the following logic diagram (Figure 5.1).

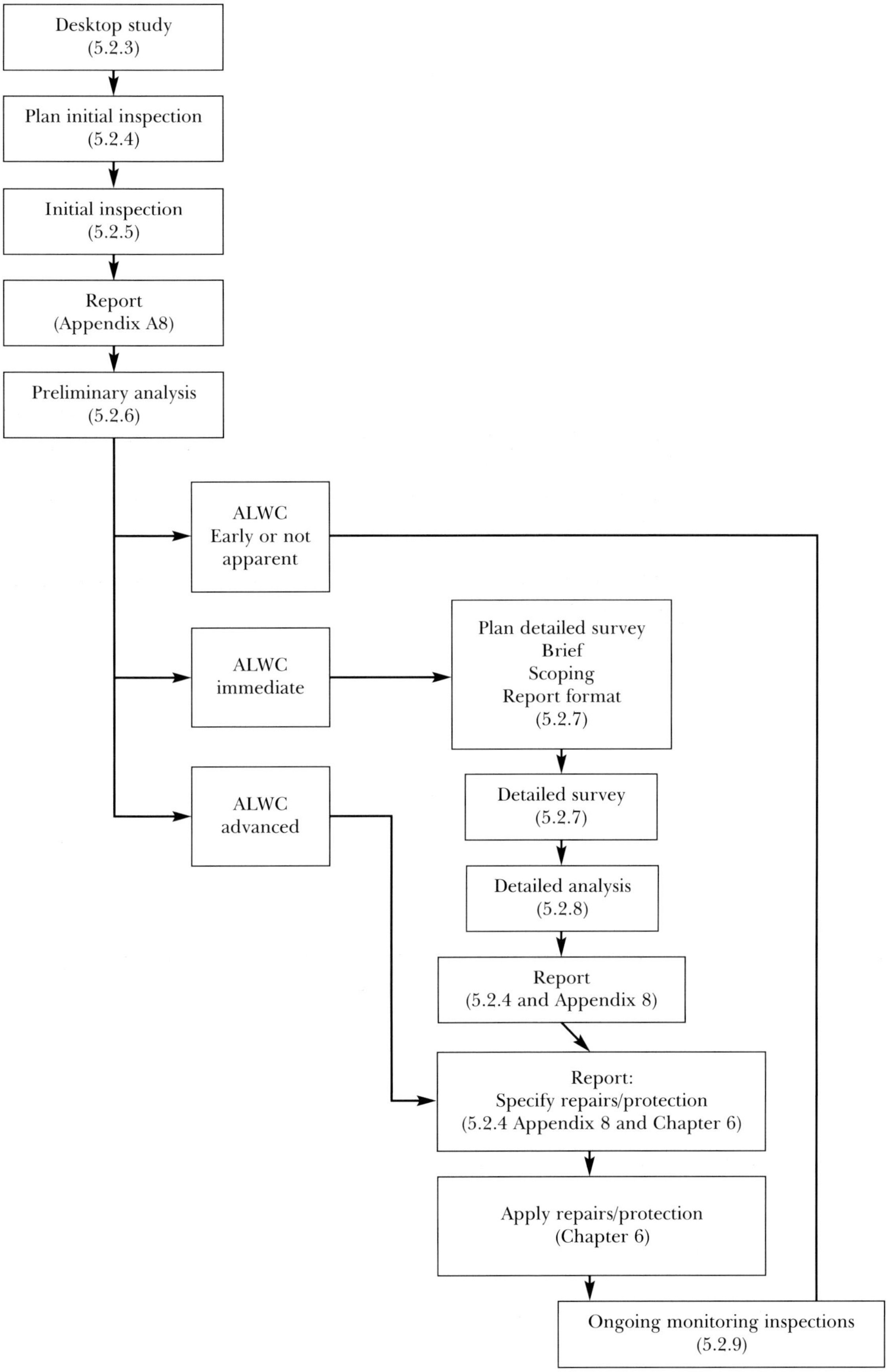

Figure 5.1 **Logic diagram for condition appraisal of a port structure**

5.2.2 Personnel

The condition appraisal should be the responsibility of an experienced chartered civil or structural engineer, with extensive knowledge of the structural analysis, design, appraisal and repair of maritime structures. A corrosion engineer, with technical competence on marine corrosion and protection methods, including for ALWC, should also be involved early on in the condition appraisal.

An inexperienced or uninformed inspector undertaking an inspection or survey of a maritime structure at higher level than the most extreme low spring tides could easily be misled by the apparent good condition of steel surfaces in these zones and be unaware of ALWC out of sight at or below the low water level (Hodgson, 1998). A selection of suitably trained inspection personnel is essential to establish a standard of quality information. It is recommended that the inspector should:

- hold specific engineering qualifications for both above and underwater inspections (see CIRIA Report 158 *Sea outfalls – inspection and diver safety*)
- be familiar with the type of structure to be inspected
- have general knowledge and experience of marine corrosion and in particular, ALWC, or have undertaken appropriate training or carried out similar investigations at other ALWC affected sites under qualified supervision prior to inspecting the structure.

5.2.3 Desktop study

Each maritime structure is different in terms of construction, access, tidal range, vulnerability and condition. A desktop study should be undertaken to:

- identify all steel fabricated maritime structures at the facility
- consider the history of the facility and the individual structures, including any previous damage and remedial works
- consider the structural form, design parameters and design methods for each structure
- identify critical areas and elements for inspection
- determine the most highly stressed elements or locations within each structure in relation to ALWC zones
- review the findings of previous inspections and surveys as appropriate
- develop a good understanding of condition
- designate a priority for inspection of structures; knowledge about the factors that may influence the incidence of ALWC (see Section 2.3.2) may also be useful in order to prioritise potentially vulnerable areas
- prepare a scope of work for the initial inspection to ensure the maximum benefit is gained.

5.2.4 Planning

It is essential that any inspection or survey is well planned, and focuses on critical elements. Maritime structures are often extensive and comprise a large number of individual elements that can only be inspected at limited specific times because of:

- tidal constraints; tidal conditions are likely to constrain the duration of any above water inspection and there may only be a few spring tides each year when such visual inspection down to LAT becomes possible. Advance planning to make the best use of these lowest of low tides for access is essential

- operational constraints; working around normal facility operations is often critical so close and ongoing liaison with operational management is essential.

Before commencing any work, a detailed brief of the scope of services required and plan of work for the survey team should be prepared. This should include the following information:

- schedule of information required
- how the inspection is to be executed
- schedule of identified hazards.

Clear priorities should be set for the inspection/survey, including:

1. Critical elements identified by the desktop study particularly those expected to be at high risk to ALWC or those already identified as suffering from ALWC from a previous inspection.
2. The scope of sampling (ie measurements/data collection) should be determined to facilitate the fast and yet relevant collection of data during the inspection/survey. The scoping should ensure that samples are not simply limited to "easy" locations or, as a consequence, a potential ALWC problem may remain undetected.
3. The priority of sampling should be determined so that the maximum benefit will be gained in the event the inspection/survey has to be curtailed eg due to deterioration in weather conditions. The information gained from the initial inspection and limited sampling will be valuable in determining the most appropriate locations and sampling procedures to be adopted in the detailed survey.

The inspection/survey reports should provide a formal record of the findings from the initial inspection or detailed survey on the structure as defined in the brief. In the case of the initial inspection, it should contain an account of the general condition of the structure at the low water level and may also include assessment of the condition in the submerged zone down to mud level. Its purpose is to identify the overall status, but primarily the worst condition, on the structure with regards to ALWC development and recommend appropriate further action to take ie on-going monitoring inspections, perform more detailed survey or install remedial measures. The detailed inspection report should include a prediction of the remaining serviceable life for the structure together with clear recommendations on the need, cost and urgency of any remedial treatment that may be required. Such remedial treatments should be aimed at reinstating structural capability and/or limiting or avoiding any further deterioration. An example of the contents of a typical survey report is given in Appendix 8.

The inspection should be subject to a risk assessment (see Sections 4.7 and 6.3). A generic risk assessment across the range of surveys is provided for guidance in Appendix 9 (Christie, 1999) from which appropriate items can either be extracted for suitable action or be added to for the development of a dedicated risk assessment, as appropriate.

5.2.5 Initial inspection

The condition of the structure at or around LAT is specifically required. The initial inspection should be a general overview, subsequently augmented by a more detailed close up inspection to be undertaken as soon as possible if any of the "tell-tale" signs of ALWC are observed. Although a number of methods of access are available (see

Appendix 7), an initial boat survey inspection at a distance 5 to 20 m away from the structure is usually adequate to provide indicative information for the presence of ALWC. It may also be prudent to sail past more closely looking carefully into the water to see, if possible, the band just below the water-line.

See Plate 4 *Close up inspection using platform and crane* on page 99 of this publication.

It is essential that this is carried out at very low spring tides when the structure is exposed at the susceptible low water zone. Subject to weather constraints, inspection may also be easier in the period between October to April when marine biofouling growth on structures would be expected to be lower. Consideration should also be given during the initial inspection to a limited underwater inspection using a diving team (see CIRIA Reports 158 and 159). The combination of boat with underwater inspection will allow visual assessment of surfaces in both the low water and continuously immersed zones for evidence of ALWC or other forms of localised attack. Clear water will improve the visual inspection but this will not be possible at some sites. Diving inspection can never be as good as inspections carried out in the dry. However, inspection at bed level may not be satisfactorily achieved in any other way.

The inspection should focus on the representative areas determined by the desktop study. A pro-forma should be designed to facilitate the recording of the relevant data. It is not necessary to record every minor defect as this would be time consuming and, perhaps, misleading.

One or more of several corrosion mechanisms could be responsible for local accelerated attack at or around LAT. Some typical visual indicators to allow an initial assessment to be made of which mechanism(s) may be operating on a given structure in the low water zone are tabulated in Table 5.1, together with some confirmatory tests which can be performed to increase the confidence level in the initial diagnosis.

Signs that should be specifically sought during the initial survey are "orange patches" (or a succession of "orange patches") in the steel around the low water zone. Discovery of such features should prompt close-up inspection for the presence of any holes and will allow removal of surface deposits at and around the identified "orange patches", visual assessment of the surface layers/substrate steel and residual steel thickness and/or pit depth measurements to be taken at these locations (see Plate 5.2). These latter measurements will enable the average corrosion rates to be estimated, as follows:

See Plate 5 *Thickness measurements being taken on steel sheet piled wall using a digital ultrasonic thickness meters* on page 100 of this publication.

$$\textbf{Corrosion rate} \text{ (mm/side/year)} = \frac{\textbf{original thickness} \text{ (mm)} - \textbf{measured thickness} \text{ (mm)}}{\textbf{service life} \text{ (years)}}$$

The original thickness values for each location may be obtained by reference to record drawings of the structure and the steel manufacturer's data sheets. If possible, thickness measurements should also be taken during the survey on sheltered areas where metal wastage appears to be minimal (eg in the atmospheric zone) in order to verify the original values at the time of construction. See Section 2.4 (Chapter 2) and Appendix 7 (Sections 3 and 5) for cautionary notes regarding possible inaccuracies in the estimation of corrosion rate.

The visual observations and magnitude of the estimated corrosion rate will establish whether an ALWC problem exists and, if so, whether it is in its:

- early stage; before repair is needed
- intermediate stage; local significant metal wastage but no holing
- advanced stage; ALWC is widespread and structural strengthening and hole repair is urgently needed.

Table 5.1 **Differences between ALWC and common corrosion mechanisms**

Mechanism	Typical indicators	Confirmatory tests
ALWC	• "classical" pattern of attack on steel sheet piling (see Section 2.3.1) • presence of "orange" patches and/or holes • surface deposit at "orange" patches can be easily removed (compared with surrounding areas) revealing black, slimy layer over shiny, pitted steel substrate • possible presence of tubercles and/or liquid filled blisters • presence of sulfide (see Appendix 7, Section 6.0) • high levels of SRB (see Appendix 7, Section 6.0).	• high rates of material loss (ie residual thickness/pit depth measurements).
Abrasion or Erosion Corrosion (high velocity water and/or entrained particulates or other water borne debris)	• damaged areas seen to be in rubbing contact with other body (eg fenders, floating pontoons, etc) • damaged areas associated with high water currents, bow thrusters, etc • polished, streamlined depressions • the attack may exhibit a directional pattern related to the direction of water flow • little or no evidence of pitting attack • absence of surface deposit or presence of "flash rusting" depending on frequency of rubbing/impact.	• surface replica; optical examination in the laboratory.
Bi-metallic corrosion	• two different metals or alloys seen to be in contact with each other with one suffering increased corrosion and the other (usually larger surface area) being partially or fully protected (eg new versus old steel, fixings welded onto structure, etc) • grooving in weld metal or HAZ at welds.	• compositional analysis on sample of metals removed from the structure.
Stray current	• isolated holes on structure • potential source of stray current or "interference" such as from impressed current CP system, railway lines, welding, etc.	• electrochemical potential survey (see Appendix 7, Section 4.0).

Representative samples of surface deposits removed from suspected ALWC patches at this time can also be taken to confirm the presence of sulfides and levels of SRB (see Appendix 7).

It is also advisable to occasionally remove surface deposit and check residual thickness at low water locations not showing any orange colouration since:

- such measurements will be useful in providing a baseline for comparison with ALWC affected areas
- the presence of ALWC can sometimes be hidden by marine growth
- other corrosion mechanisms may be occurring locally, resulting in significant metal thinning
- on a steel sheet piled wall, corrosion may occur from the rear face although this is considered to be rare unless a hole has already formed in the piling wall.

It is recommended that only reasonably "mild" techniques are used in order to avoid removing well adhered surface deposits. Cleaning the "orange" deposits by hand from the steel surface at ALWC patches should prove to be relatively straightforward due to the low adhesion of these deposits. Cathodic cleaning, under controlled application conditions, has also been demonstrated to locally remove these less adherent deposits as part of the first stage of a new ALWC treatment (see Section 9.2). The hydrogen generated on the steel surface (as a result of local cathodic polarisation of the structure) is used to remove surface deposits just at the ALWC patches leaving the deposits in surrounding areas intact (Marshall, 2002). Removal of such deposits at ALWC patches is advantageous since they are known to be locally aggressive (ie sulfide rich, acidic, etc) to the substrate steel and their removal may be expected to reduce the overall rate of metal loss. As a general rule, removal of the more adherent surface deposits surrounding the ALWC patches should be kept to a minimum. This is because these more adherent forms of surface deposit in the low water and submerged zones may be providing some degree of natural protection against corrosion to the substrate steel. Care should be taken not to reduce or destroy this effect. This can be checked by measuring residual thickness at a limited number of locations and comparing the calculated corrosion rates obtained with the "traditional" general (uniform) corrosion rates for this zone given in Table 2.1.

It is important to note that surface deposits in areas adjacent to ALWC patches in the low water zone can also show the same layered structure with visual features similar to those described for ALWC above. However, there are important differences compared with ALWC patches. For these adjacent areas:

- the orange/black surface deposit could be covered with an intact outer marine fouling layer
- the corrosion will be more general (uniform) in nature and the severity of attack will be significantly less
- the adhesion of the surface deposit will be significantly greater than at ALWC patches, but not as well adhered as "traditional" rust scale (ie hydrated iron oxide deposits) seen higher up the structure
- the steel surface beneath the surface deposit, although clean, will appear "dull" (ie a matt finish) rather than "shiny".

It is important to identify and accurately record the exact locations of the ALWC patches or holes using a survey drawing with grid system (see Figure 5.2). A photographic survey, before and after cleaning of non-adherent areas, can also be used to give a quick and indicative record of attack. Long panoramic photographs produced digitally and printed in banner mode can be created. Wherever possible all reference points should be located above water and items of interest located by suitable horizontal and vertical controls. For a steel piled structure this would typically involve identifying individual piles (ideally, using the original construction numbering system to facilitate the tracing of repairs) with a unique reference mark (eg using permanent yellow line spray marker paint or similar, marking inpans as even numbers and outpans as odd numbers) and the use of a suitable depth positioning aid, such as a calibrated datum line or depth gauge. The positions of the confirmed ALWC patches and/or holes are then marked up on a survey drawing to allow quick identification for any remedial works required or subsequent inspections and/or measurements at the same point. The location of ALWC patches will be reflected in the results as isolated low thickness readings among a group of surrounding satisfactory readings at or around the LAT zone. To facilitate visual identification of thinned areas on the survey drawing, a colour coding system may be adopted using different colours to correspond with different ranges of thickness loss.

Orientation around tubular piles should also be recorded using "clock positions" making sure to identify the seaward facing position, for example. In the case of an underwater inspection, other more sophisticated positioning systems such as ADAN acoustic navigation, differential global positioning or similar systems are available to identify a diver's location.

For solid structures, such as steel sheet piled walls retaining fill, a walk-over survey should also be undertaken to note any apparent settlement that may have been caused by loss of fill through holes in the sheet piles. Water ponding on pavement surfaces after heavy rainfall may also be indicative of loss of fill. A walk-over survey is particularly necessary for older structures (more than 20 years old) that may not have been specifically inspected for ALWC in the critical zone before. Surface drilling can also be used to identify voids not apparent from the surface.

5.2.6 Preliminary analysis

Experience indicates that for each hole discovered during a preliminary ALWC inspection or survey, several others may be revealed during subsequent remedial work. The extent to which this is true clearly depends on the level of inspection that has been undertaken. Nevertheless, this should be taken into account when inspection reports are being evaluated.

The preliminary analysis will indicate which of three options should be pursued:

1. If there is no evidence of ALWC or it is detected in an early stage (ie before structural strengthening is needed), then initiate a programme of regular monitoring.
2. If ALWC is detected in an intermediate stage (ie examples of areas showing local significant metal wastage but no holing) then undertake a more detailed survey while planning, preparing and budgeting for the necessary remedial work.
3. If the structure is seriously affected by ALWC, so that sections are found to be in an advanced stage with local areas showing serious metal wastage and/or holing in the low water zone, then undertake an urgent remedial programme. Experience suggests that there is likely to be similar damage elsewhere on the structure in the low water zone. It may be more prudent to undertake a more detailed survey as part of the remedial programme.

5.2.7 Detailed survey

A number of methods of access are available for undertaking a detailed survey, as described in Appendix 7 (Section 1). As with the initial inspection, any higher area already surveyed should not be taken as indicative of the condition of the structure in the continuously immersed zone. This area should be examined by a diving team with cameras or more thoroughly by use of a limpet dam. The specific circumstances of the site will determine how the survey should be carried out. The principal factors to be considered include:

- safety
- time available
- quantity and quality of measurements required
- economics.

Sheet pile type – LARSSEN 5 d=22.1mm , t=11.9mm

Quay level +6.3m

Residual steel thicknesses recorded in mm

Pile references to outpans

Underside cope level +5.3 m ACD75

Areas displaying corrosion products

50-70% orig. thickness

less than 50% orig. thickness

Pile reference:

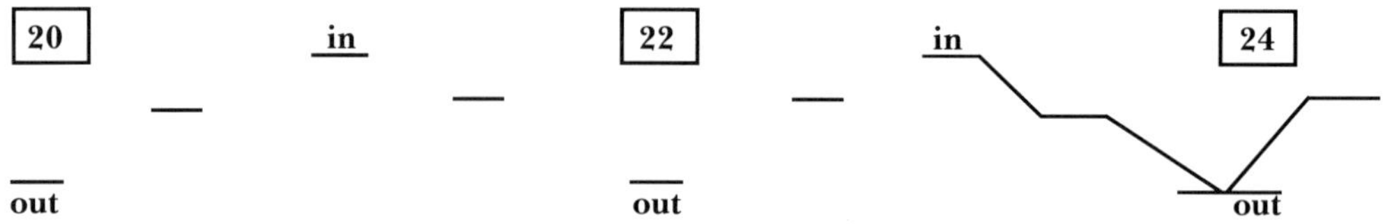

Level to ACD 75 (m)	Dist. below top cope (m)	20			in			22			in			24		
5	**1.3**															
4	2.3	11.6	21.6	11.9	11.3	20.6	10.2	9.9	21.6	10.7	10.5	21.6	10.3	10.9	20.1	11.0
3	**3.3**	11.7	21.7	11.8	11.3	22.0	10.2	10.8	22.0	11.3	9.9	21.6	11.2	10.9	20.0	11.3
2	**4.3**	11.5	21.0	11.6	10.7	20.4	11.7	11.0	19.8	11.0	11.5	20.6	10.5	9.8	19.6	10.3
1	**5.3**		14.3	10.5	10.2	17.4	10.8	10.7	18.7	10.5	10.0	20.3	10.2	10.5	18.7	10.2
											9.8					
0	**6.3**	11.5	20.8	12.0	10.4	19.9	9.6	11.4	7.5	10.4	10.2	13.1	10.6	11.8	20.9	10.5
											9.1					
-1	**7.3**	11.3	21.7	11.9	10.2	19.6	10.2	11.0	13.0	11.4	10.6	18.9	11.1	11.6	20.7	11.9
-2	**8.3**	11.2	21.1	11.5	11.3	21.9	9.3	11.3	21.2	11.4	11.3	21.9	11.7	11.3	21.0	11.9
-3	**9.3**	11.4	19.8	11.9	10.9	21.7	9.8	11.9	17.7	11.8	11.5	21.0	11.2	11.3	20.7	10.9
-4	**10.3**	11.0	21.4	11.0	11.2	22.0	11.6	11.4	21.2	11.8	11.3	21.5	11.9	11.6	17.8	11.3
-5	**11.3**	11.3	21.7	12.0	11.5	21.3	12.3	11.8	21.7	11.9	11.3	21.5	11.5	11.9	21.0	11.8
-6	**12.3**	Bed	Bed	Bed	Bed	Bed	Bed	Bed	Bed	Bed	Bed	Bed	Bed	Bed	Bed	Bed

Notes on pile appearance, holes, features, etc:

No holes evident

Severe pitting in areas marked

Plating recommended pile 22 out-pan

Figure 5.2 Example of typical survey drawing for steel sheet piled wall

The target should be to inspect in the dry as far as possible, to improve the quantity of data. However, quality should be unaffected if competent divers are used and the visibility is good.

Nevertheless, if holing has already occurred then urgency may outweigh economy and an appropriate access/inspection method should be adopted to minimise delay, having due regard to safety.

The detailed survey of a structure known to be affected by ALWC should have clearly defined priorities, developed from the initial inspection and analysis. Within each area of the structure, the survey will often be required to identify:

- the worst areas of general thinning (location, extent and size)
- the presence and depth of corrosion pits at these thinned areas (maximum localised attack)
- the representative thickness of the remaining sound metal for structural purposes.

There may be few or numerous individual steel components to check for evidence of ALWC, all of which may be under a variable level of attack. In practice, the actual number checked is often some arbitrary value for example, in the case of a sheet piled wall, typically 10 per cent of the piles (Steele, 2002). However, where the size of the structure is extensive, consideration should be given to applying a statistical approach to determine the minimum number and the distribution to inspect in order to give confidence in results on its overall condition. The number of relevant samples to be acquired will depend not only on the total number of structures that are to be inspected, but also on their representative location, for example:

- orientation
- sunlight versus shade
- exposed to or sheltered from waves, current, prop wash etc
- low versus high pollution site.

Other data, in addition to thickness measurements, may also be considered for the detailed survey, as follows:

- an electrochemical potential survey across the structure while under immersed conditions (ie when the tide is in) using a diving team. The purpose would be to determine the presence of any macro-corrosion cells that may be in operation either horizontally or vertically on the structure (see Appendix 7, Section 4). This would assist in deciding on the cause(s) of the problem producing the localised attack and hence the most appropriate protection method for the structure and how it is to be applied. The diver would be instructed to take structure to seawater potential readings using a hand unit prior to removal of any surface deposit, using the same grid identification pattern as for the thickness measurements. In practice, any regions on the structure showing variations greater than 20 mV should be compared with those regions subsequently identified with accelerated thinning of wall thickness to check for possible correlations. A positive correlation would require further investigation in order to determine the cause (eg oxygen concentration cell, etc)
- a limited number of small diameter cores (no less than 25 mm diameter) taken from the steel surface using a trepanning tool in order to allow laboratory determination of steel composition at different locations, particularly when structures are of different ages or bimetallic corrosion is suspected. Where possible, such cores

should be taken in the atmospheric zone to facilitate subsequent repair

- samples of the surface deposits taken for subsequent laboratory analysis. However, certain precautions will be necessary to ensure that the samples are not spoilt during sampling, storage and transit (see Appendix 7, Section 6).

5.2.8 Detailed analysis

The structure should be analysed by a competent design engineer, either to prove the adequacy of the structure and assist in determining the remaining serviceable life without protection, or to determine the extent of necessary repairs. Fortunately, structural failures are extremely rare and sometimes do not occur even when the standard analytical techniques suggest they should have. Nevertheless, it is essential for the owner to be confident that the structure is operating safely and able to carry the loads that will be imposed on it. A potential reduction in the load-carrying ability of the structure will be indicated by loss of section, or holing.

a) Requirement for structural analysis

A severely degraded structure, considered unstable by structural analysis, sometimes continues to stand.

Typical factors that can contribute to the unexpected apparent stability of such a structure include:

1. Deflection of a failing sheet pile wall will temporarily reduce the loadings on it from the contained fill material, which may have a measure of temporary stability of its own.
2. As members start to fail and deflect they transfer excess load to adjacent members, for example decreasing strength of a tied steel sheet pile wall will cause it to deflect, thus transferring more load into the tie rods and anchorages.
3. Loss of retained fill material through holes in the retaining wall will reduce the loading on it.
4. Using up factors of safety.
5. It is also highly likely that the structure will not have been acted on by the most severe loads, or combination of loads, during its life to date.

However, it should be stressed that these beneficial factors can not be relied upon when deciding whether a structure is safe to use, because they can be eroded suddenly and catastrophically. Therefore, a degraded structure that remains standing without apparent distress should not be considered sufficient reason to neglect undertaking a structural assessment, nor to justify its continued use.

b) Methods of design and analysis

In the UK British Standards, and more latterly Eurocodes, have provided guidance on parameters and methods of design of maritime structures.

Design methods have changed considerably in recent years, principally as follows:

- computer aided design enables more intensive analysis to be undertaken, eg assessment of the soil/structure interface in the consideration of steel sheet piled structures
- limit state design methods allow safety factors to be tailored to particular items rather than applied globally to all loads.

Other changes have included:

- development in methods of analysis of steel sheet piled retaining structures
- differing guidance on rates of general corrosion (ie not allowing for ALWC)
- introduction of 2-D and 3-D finite element and difference analysis of retaining structures.

Improved analysis has made structures less conservative by enabling, among other things, the removal of some redundancies in the design, offsetting rising construction costs. Increased adoption of design-build contracts, where the contractor can win the job because of the economy of design, has reinforced this trend.

Hence, modern, more economical designs will have less spare capacity. It is, therefore, even more critical that the possible effects of ALWC are fully considered in the design and the appropriate protective measures applied.

The re-analysis of an affected structure using more modern methods of design and analysis may reveal additional reserves of load capacity which were not apparent during the original analysis. It may avoid, or at least reduce, the extent of repair work that might otherwise be required, or allow protection works to be delayed.

The level and reliability of the re-analysis will depend on the data available; "as built" records will prove particularly valuable if they are accurate.

c) Confirmation of structural adequacy

A more rigorous structural analysis may be able to prove the adequacy of the structure for the loads that will be imposed on it. The following may be considered to identify whether this is the case:

Design issues

1. The original design methods may not have been the most rigorous and may, therefore, contain some conservatism.
2. The standard section used by the designer may have spare capacity over the strict design requirements.
3. Where a constant section designed to predominantly resist bending stresses is used over the full height of the member, it will have reserve capacity over much of its length. Care should be taken to ensure that the load case giving the most onerous stresses at the most corroded section is selected by the designer or the assessor.
4. Where a standard section has been used over the full length of the structure or a discrete length of it, there may be spare capacity where the conditions are less onerous (eg decreasing height of a wall along its length).
5. Soil parameters may have been conservatively assessed, perhaps because of lack of information available at the time the initial design was undertaken.

Construction issues

1. The pile section installed may have been larger than required by the design to allow for temporary conditions during construction, for example driving requirements.
2. The pile section installed may have been larger or of higher grade than required because of availability of material (eg surplus gas pipe as piling material).
3. The piles may have been driven to greater (or less) penetration than specified.

Changed use

1. Allowances for future loadings and changes of use (such as a greater dredged depth in front of the wall) may not have materialised.
2. The facility might be operating at reduced loading capacity.

However, it should be noted that some factors not considered in the original design, may decrease the adequacy of the structure. These might include:

1. Difficulties encountered during construction. These may already have utilised any spare capacity of the designed structure.
2. Additional loadings may have been imposed.
3. Further dredging may have been carried out or scour may have occurred.

d) Other factors

- change of service – there may be scope for the owner to modify use of the structure, for example imposing load restrictions or relocating bollards. It should be noted, however, that in most sheet pile walls the predominant loading which causes bending comes from the retained fill material
- design codes – the design code used may be less conservative than others
- owner's design codes – the owner may, from experience, consider adopting higher stresses under particular load combinations. Some owners, such as large petrochemical companies, produce their own design requirements which may be less onerous under particular circumstances than the standard requirements.

e) Types of structure and loadings

Axial forces are usually fairly constant along the length of a member, whereas bending forces will vary considerably and can reverse with level and direction of application of load. Horizontal impact (dynamic) loading from berthing, accidents, ice flows, bollard pulls etc can also occur.

Table 5.2 identifies the predominant stresses on typical supporting elements for various types of structure.

Table 5.2 **Predominant stresses on elements for different types of structure**

Type of structure	Typical element	Predominant stress
Rigid piled skeletal structure	Tubular or box pile, wall thickness can vary	Axial/bending
Portal frame structure	Tubular or box pile	Bending/axial
Sway structure, eg monopile dolphin	Tubular piles, varying wall thickness	Bending
Containment structure supporting high vertical loads, eg sheet pile wall also supporting crane rail	Sheet pile, king pile, combi pile	Bending and axial
Other containment structure, eg sheet pile wall	Sheet pile, king pile, combi pile	Bending

5.2.9 Ongoing monitoring inspections

A programme of regular monitoring inspections, to record the condition of the structure and/or the effectiveness of any previously installed remedial measures, will allow any required maintenance to be effectively and economically implemented (PIANC, 2004). Inspection intervals should be reviewed following each inspection but should not exceed two years depending on the age of the structure or every six months where there is already a cause for concern. Unprotected, or inadequately protected, structures over 20 years old are probably at greater risk of showing ALWC in an advanced stage compared with newer structures.

The monitoring programme should form part of the asset management strategy (see Chapter 4) and should take into account the age and type of structure together with any environmental or safety risks specific to the site. The aim of the programme should be to monitor for on-going deterioration by ALWC and to facilitate planning, preparation and budgeting for maintenance work. Emphasis should be placed on elements identified by the desk study to be critical or high-risk components.

Regular inspection at previously identified or suspected ALWC sites allows the loss of section at these locations to be monitored. Once the thickness falls below some critical value, determined by structural analysis, then maintenance will become necessary to maintain structural integrity and/or to prevent holing, ie prior to the ultimate limit state failure (see Figure 4.4).

Predictive tools can be used to aid the determination of an intervention date for maintenance. For modelling purposes a confirmation of the presence of ALWC is needed initially and also the remaining section thickness. Then, a simple spreadsheet model can be set-up to predict the intervention date using measured rates of deterioration on the actual structure in question and/or estimates of ALWC corrosion rates given in this guidance (see Table 2.1). It is important that the corrosion rates used in the model have been obtained from ALWC sites and are not general (uniform) corrosion rates from adjacent areas as these will give a distorted picture and not accurately predict the intervention dates.

Summary

- competent inspection is required for the identification of ALWC
- regular inspections followed by early remedial measures are recommended. The sooner ALWC is detected and addressed the cheaper the remedial solutions will be
- the quality of inspection data that can be obtained is usually a function of cost
- it is essential to carry out an initial inspection for ALWC at or around LAT and, if present, establish whether it is in an early, intermediate or advance stage. This will allow any maintenance and protection to be prioritised and budgetary considerations to be made sooner rather than later
- only reasonably "mild" surface cleaning techniques should be used during inspection in order to avoid removing well adhered deposits which are probably providing a degree of natural protection against corrosion to the substrate steel
- it should be recognised that ALWC may also occur in the continuously immersed zone and, therefore, this area should also be inspected
- depending on the findings the following is recommended:
    - if ALWC is absent or in its early stage (ie before structural strengthening is needed) then initiate a program of ongoing monitoring at regular intervals, usually not exceeding two years. The "do nothing" option is only valid if the estimated remaining life of the structure is sufficient for requirements, or future construction plans suit the current rate of corrosion. Once holes have formed repairs will become more difficult and expensive
    - if ALWC is in an intermediate stage (ie local significant metal wastage but no holing) then carry out a more detailed survey while planning, preparing and budgeting for the necessary remedial work. Consideration should be given to adding extra steel to identified "thinned" areas before holing occurs and/or implementing one or more of the protection methods (see Chapter 6) before it deteriorates into an advanced stage
    - if ALWC is in an advanced stage (ie local areas showing serious metal wastage and/or holing) then undertake a remedial programme (ie structural strengthening and/or hole repair) on an urgent basis. Since there is likely to be similar damage elsewhere on the structure in the low water zone, it may be prudent to undertake a more detailed survey as part of this remedial programme. Repaired or replaced sections should be protected against future attack (see Chapter 6) – the risk of ALWC reoccurring at the same location would be considered high
    - monitor any further deterioration of the structure and installed remedial solutions by regular, ongoing inspections at suitable intervals (see Chapter 8).

6 Repair and protection

The extent to which the structure has been affected by ALWC and the action required to ensure it remains operational for the required service life will be determined by the inspections, surveys and structural analysis carried out (see Chapter 5) together with, as appropriate, the requirements of an asset management strategy (see Chapter 4).

This chapter considers the development of a strategy for making good and protecting structures affected by ALWC and indeed any other corrosion mechanism that produces rapid localised attack on a structure. Methods of repair are considered, including the corrosion implications that should be avoided and methods of access to undertake the work. Finally, methods of protecting existing ALWC affected structures (including any repairs) and of new build structures are described. There are a number of case studies included in this chapter (and additionally in Appendix 11) to give practical advice on what has been done to repair and protect structures in certain applications. However, it should be emphasised that these sites have not been visited as part of this study nor have the remedial measures been confirmed as achieving the required objectives. Risk assessment should be carried out both to determine when action is to be taken and to ensure mitigation measures are performed in a safe manner.

6.1 Selection of strategy for repair

Once corrosion rates have been calculated and analysis of the critical thickness/damage has been performed, then the remaining effective life of the structure can be estimated. If ALWC is active, then this prediction will be significantly less than the intended design life. Measures will have to be put in place to ensure that structural integrity and operational ability are maintained.

Depending on the severity of corrosion, this will involve:

- protection against future corrosion using measures described in Section 6.5
- repairs, or modified use, as described in Sections 6.2 and 5.2.8 respectively, followed by protection against future corrosion using measures described in Section 6.5
- reconstruction of part or all of the structure.

Each combination of site and installation is unique in terms of environment, design, construction and operation. Furthermore, as described in Chapter 5, the point of deterioration at which a particular maritime structure will require repair and/or protection will be unique to each structure (or even each structural member) and will depend on the required and remaining structural capacity and the rate of corrosion. The design of repairs and protection, therefore, requires the involvement of both structural and corrosion engineers with experience of maritime facilities.

Selection of the correct strategy will principally be determined by:

- the economics and availability of funding; some work may have to be deferred subject to ensuring that safety and structural capacity are not compromised
- the shortfall between the required and predicted remaining life of the facility
- any reduction that may have to be imposed on operational loads

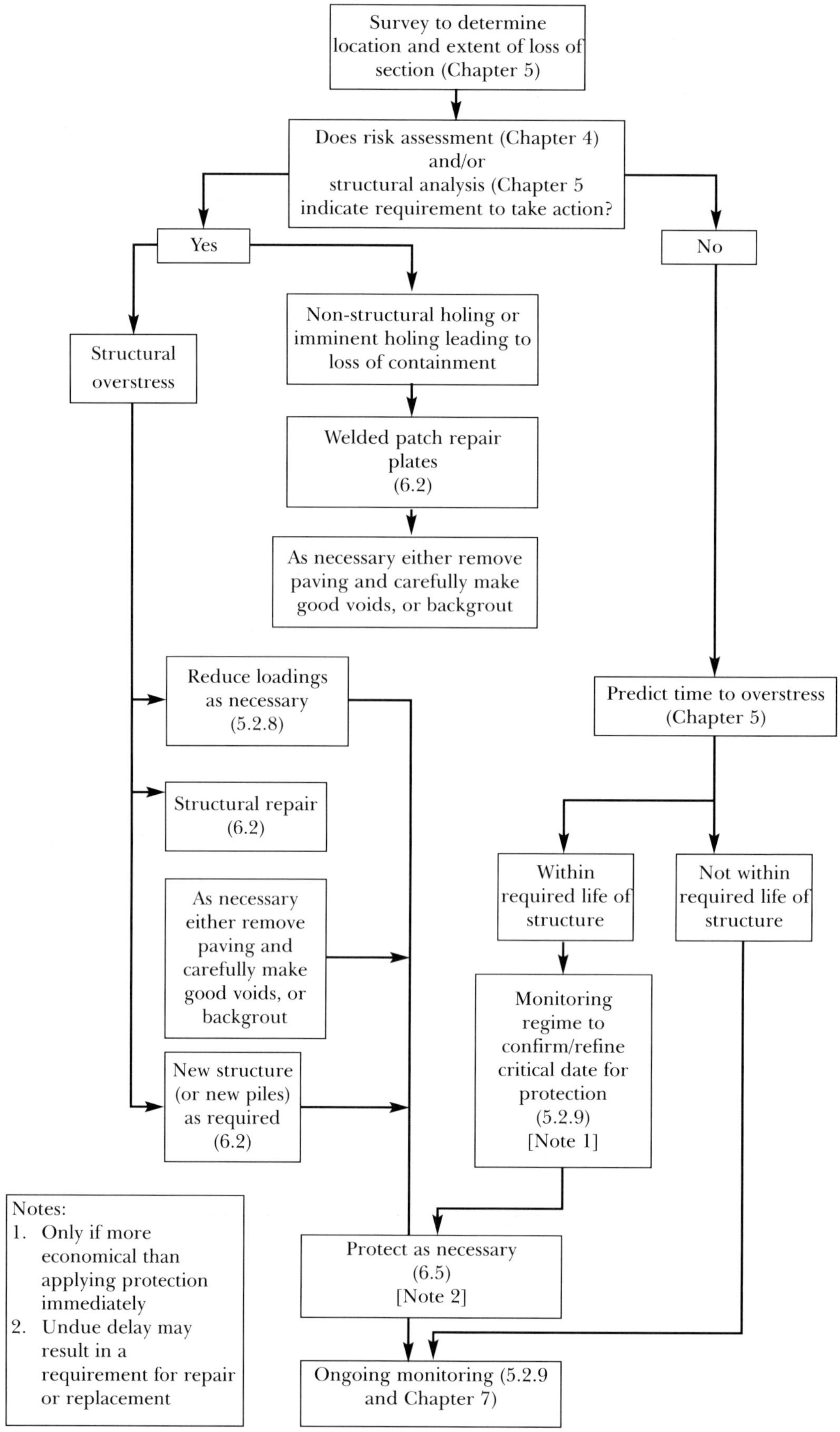

Figure 6.1 Logic diagram for repair of skeletal and solid structures

- the usage of the facility
- the cost of repair and/or protection; if steel section loss is severe and/or if holes have become unmanageably numerous or large then repairs can become uneconomical
- selection of the most cost-effective solution and resources
- whether access to the damaged areas can be achieved safely and practicably
- future maintenance requirements
- rapid versus cheaper methods.

Figure 6.1 provides a typical logic sequence for selection of a strategy for both skeletal and solid structures.

6.2 Repair techniques for ALWC

A summary of the structural repair techniques, available for both skeletal and solid structures and their suitability for structural purposes, is given in Table 6.1. These repair techniques, which are also suitable where other forms of rapid, localised corrosion may be present, are described in more detail in Sections 6.2.1 and 6.2.2 respectively.

Table 6.1 Summary of structural repair techniques

Repair technique	Skeletal structures	Solid structures
Welded patch repair plate	×	×
Welded strengthening plate	√	√
Plating plus reinforced concrete infill	NA	√
Reinforced concrete collar / encasement	√	NA
Concrete pile infill	√	NA
Reinforced concrete facing	NA	√
Spliced in replacement length of pile	√	NP
Replacement pile or steel sheet pile	√	NP

Key:
× Not suitable for structural repair
π Suitable for structural repair
NA Not applicable
NP Not (normally) practical

6.2.1 Skeletal Structures

Welded patch repair plates – used for non-structural repairs to make good localised thinning or holing to prevent ingress of aerated water leading to increased corrosion. They may also be used to add extra steel to vulnerable areas before holing occurs. The procedure typically involves:

- preparing a repair plate (typically 2 m minimum height) shaped to match the cross-section of the pile and of suitable thickness (10 mm minimum) to allow welding to sound parent metal

- cleaning the area to be plated using high pressure water jetting, 8 × 107 N/mm² (ie 800 bar) or similar suitable system
- pumping out any water within and around the pile sufficient to allow welding
- fixing the plate over the area affected by ALWC and welding around the perimeter.

Accelerated (ie bimetallic) corrosion should be expected when connecting a relatively small, new steel plate to a large area of old steel, unless suitable protection measures are used (see Section 6.4). Incomplete or "tack" welding around the perimeter to connect the repair plate and the substrate steel is not recommended since this may result in crevice corrosion although the possibility of this occurring for steel in seawater is not certain. If it is not possible to guarantee a continuous weld then an epoxy based paste or cement grout can be injected into any crevices or voids remaining after welding the plate to the substrate steel. This is required to ensure that the repair is completely sealed. Alternatively, a layer of non-shrink epoxy based paste or cement grout can be applied to the back surface of the repair plate before fixing.

Box 6.1 Case study for welded patch repair plate on skeletal structure

A limpet dam was fitted around each Rendhex pile in turn and the seawater pumped out. The limpet dam enabled the pile to be cleaned, surveyed and repaired in a dry environment. A pair of 15 mm Grade 270, 2 m long plates (bent to suit shape of the Rendhex) were offered up, one to each side of the pile. These plates were positioned so that they spanned ALWC holes and thinned areas, and were welded to the pile using MIG welding techniques. Welding was a continuous 10 mm fillet all round back to the pile. The finished result was a 2 m high sleeve spanning the ALWC susceptible zone.

Thinned corroded surfaces will be irregular and welding to them can be difficult. Care should be taken to ensure that steel plates span over such areas and the welding is carried out to sufficiently thick and sound sections of the substrate steel otherwise holing can result. It is often useful to try and weld to clutches.

Welded strengthening plates – used for structural repairs for more extensively corroded areas comprising a steel plate shaped to the cross-sectional profile of the pile and welded to sound steel. The length of the repair plate, determined by structural considerations, comprises:

- a length above the hole or weakened area to transfer loads from the pile into the plate
- the length to span the hole or weakened area
- a length below hole or weakened area to transfer loads back into the pile.

The procedure is similar to that used for the welded patch repair plate above; except that the plate will be larger and will incorporate structural welds able to transfer the loads from the parent pile to the repair plate and then back to the parent pile. Good quality welds extending the full perimeter of the plate, with no voids or crevices are preferred although the possibility of crevice corrosion resulting if this is not done is uncertain (see Section 6.4). Often it is necessary to incorporate "weld windows" into the plate to permit sufficient stress transfer where loads are large and there is insufficient plate perimeter without them.

Preparation of such plates can be difficult because of the cross-sectional profile of the pile. The maximum thickness of the plate will be determined by ability to bend the plate to the required profile. The material used for the plate and welding consumables should be carefully chosen (see Section 6.4) to reduce the risk or impact of bimetallic corrosion and/or preferential weld corrosion unless suitable protection measures are used (see Section 6.4). Significant cold forming in preparation of the plates may also increase their susceptibility to corrosion.

Box 6.2 Case study for welded strengthening plate on skeletal structure:

A repair system comprising two half shells rolled from 15 mm mild steel plate, 1 m long was used to repair tubular piles. The piles were of spiral weld construction so the welds initially had to be ground flush with the parent metal to allow the shells close contact with the pile wall. To position each half shell at the correct height, a simple clamp arrangement was placed around the pile above the working area. The shell was hung from the clamp using a ratchet block, which allowed the diver to move each shell into the required position for underwater welding.

In-situ, reinforced concrete collar or encasement – piles can be encased with concrete or grout, typically formed using a (permanent) grout bag, fabric bags, fibreglass formwork or steel shuttering (temporary or permanent).

The concrete should be of suitable quality and of sufficient cover to minimise the rate of chloride penetration reaching the reinforcement. The use of cathodic protection or more corrosion resistant reinforcement may also be considered to overcome or further reduce the risk of subsequent chloride attack. BS6349: Part 1 provides guidance on the design and specification of concrete mixes for the maritime environment. Specific additives may also be included in the mix in order to obtain high early strength, to enhance workability without segregation or to provide some degree of corrosion protection to the steel reinforcement. Polypropylene fibres may also be introduced into the concrete mix to reduce the risk of cracking.

Box 6.3 Case study for pile collaring system on skeletal structure

A pile collaring system has been used for the repair of Z and U box piles and tubular piles affected by ALWC in various locations around Ireland and in Shetland. In all cases ALWC was in an advanced stage with a high percentage of piles exhibiting holes or low residual steel thicknesses. The pile collaring system is designed to protect the piles against future corrosion and enhance the structural strength of each pile restoring design factors of safety against bending moment. It has been used as a cost-effective alternative to large scale demolition and reconstruction of the structures.

The pile collaring technique is a diver intensive operation and involves:

1. *Pre-cleaning of the existing piles using high pressure water jetting to remove all marine growth and corrosion product; this is to ensure adequate bond between the pile and the high strength grout.*
2. *Installing a structural galvanised steel collar (fabricated in two half shells) around the pile from below seabed to the concrete soffit or just below, depending on access and the existing condition of the piles. A zinc rich paint system is used to coat any remaining exposed steel under the soffit.*
3. *Plugging the base of the collar and filling the annulus with high strength, non-shrink grout.*

Where the piles had developed holes these were either filled with concrete or the hole patched prior to filling the pile/collar annulus with grout. The completed repair relies on the pile/grout bond and shear studs welded to the inside of the collar to transfer the load from the existing pile at the top back into the pile at seabed.

The effect of increased rigidity of the element as a result of concrete encasement and/or the possibility of vessel impact, wave action etc can cause cracking of the concrete. This may allow direct ingress of seawater to steel reinforcement and/or steel substrates which, if inadequately protected, can result in spalling of the concrete due to the ensuing corrosion. See Section 6.4 and Appendix 10, Section 3 for further cautionary notes regarding corrosion implications of concrete based repairs.

Concrete pile infill – concrete or a suitable grout can be pumped into the void within a box or tubular pile to provide strengthening. While the repair will assist in preventing the pile wall from buckling, the reliable transfer of loads between the wall of the pile and the concrete or grout can be a problem. A case study involving the use of reinforced concrete infill technique in conjunction with a wrapping system is given in Appendix 11 (ALWC, Case study 4).

Splicing – it may be possible, depending on the temporary structural implications, to cut out a small damaged section of pile and splice in a replacement length. However, this is potentially dangerous and will be difficult to achieve in practice. It also has corrosion implications similar to those for repair plates (see Section 6.4).

Replacement pile – where piles are beyond economical repair and the existing structure permits, an additional pile can be driven adjacent to or in place of the affected pile and tied into the existing structure. Removing a damaged pile and slotting a new pile into place in an existing structure is difficult and likely to be impossible in many situations. In practice, the concept of using a replacement pile can only realistically be used if a new wall alignment is created.

Box 6.4 Case study for replacement pile system on skeletal structure

A 914 mm diameter tubular raking pile supporting a mooring dolphin at a terminal in Orkney was damaged beyond repair at low level by vessel impact. The chosen repair method comprised:

- *cutting off the pile as low as possible using divers*
- *forming a hole through the reinforced concrete cap using ultra high pressure water jetting*
- *pitching and driving a new pile through the hole and on a steeper rake to avoid the remains of the old pile*
- *anchoring the new pile into bedrock in a similar manner to the original pile*
- *making good the cap.*

6.2.2 Solid structures

Welded patch repair plates – used for non-structural repairs to make good localised thinning or holing and to prevent or halt the loss of backfill from behind the wall. May also be used to add extra steel to vulnerable areas before holing occurs. A typical procedure involves:

- preparing a repair plate shaped to match the cross-section of the pile and of suitable thickness and size to allow welding to sound parent metal
- cleaning the area to be plated using high pressure water jetting, or similar
- filling any holes with fast-setting cementitious mortar or expanding foam or similar to prevent water passing through the wall and affecting the welding operation. While having little structural strength, rapid curing mortars have also been used underwater by divers to temporarily fill (by hand) small diameter holes in steel structures. They are typically supplied in a dry form and mixed on site with seawater to a stiff consistency. Because of the rapid curing it is possible to build up

several layers to fill deep holes. No temporary seal should stand proud of the surface because it would prevent subsequent plate fixing

- fixing the plate over the area affected by ALWC and welding around the perimeter.

The plate used should be of sufficient thickness and the fixing method of sufficient strength to resist the loads that will be applied to the plated area by the fill. The maximum thickness will be determined by the ability to bend corners to the required profile (for mild steel, the radius is equal to two times the plate thickness).

An alternative to welding of the repair plates is mechanical fixing using nuts and bolts. The threaded studs are first connected to the structure by either conventional or friction welding. Friction stud welding can be used to weld fasteners on the normal face both above and below water with minimal or no surface preparation. Alternatively, the studs can be fired into the steel using a bolt gun with due regard to the safety implications. The bolt gun can be used both above and below water to give a rapid means of fixing. The bolts should be fired into solid sections of a pile and plate sizes adjusted to suit the bolt locations. If the pile thickness is not sufficient, then the result will be either a poor fixing or the bolt passing right through the pile. The use of bolt guns should normally only be implemented in an emergency since the quality of the finished repair is always suspect.

As for skeletal structures, bi-metallic corrosion (due to connecting new to old steel) and possibly crevice corrosion (eg where the perimeter edge of the plate is not completely sealed or between the nuts/ bolts and the plates) is to be expected if suitable protection measures are not taken (see Sections 6.2.1 and 6.4).

Box 6.5 Case study for welded patch repair plate on solid structure

The structure was a 30 year old sealing dam constructed from flat pan Frodingham piles. The piles were holing at 0.5 mCD thus jeopardising the integrity of the gravity cellular cofferdam. Elsewhere on the structure, there was no significant corrosion. Two types of limpet dam were used to provide a dry environment from within which the piles could be cleaned, inspected and repaired. The first type fitted the convex shapes of the cells and infill arcs. The second fitted the cleavage areas between the main cells and adjacent arcs. Grade 270 12 mm plates were chosen to repair all piles at the ALWC susceptible zone. Plates were continuously welded all round (8 mm fillet) using a MIG welding technique.

Welded strengthening plates – structural repair used for more extensively corroded areas comprising a profiled steel plate welded to sound steel; U-shaped plates are used to enclose damaged outpans (see Plate 6.2) while other profiled plates can be used to suit specific areas of damage (see Appendix 11; case studies 1 and 2). The length of the repair plate will be determined by structural considerations, as given for skeletal structures.

See Plate 6 *Repair using welded u shaped steel* on page 101.

The procedure is generally similar to that for the patch repair plate described above; except that the plate will be larger and will incorporate welds able to transfer the structural loads. As with skeletal structures, it may be necessary to incorporate "weld windows" cut into the plate.

Preparation of the profiled repair plates can be difficult to achieve, particularly for piles that have suffered distortion as a result of driving, deflection under load or berthing damage. As for skeletal structures, the material used for the repair plate and the welding consumables should be carefully chosen (see Section 6.4) and the weld should be continuous.

Often the existing clutches are used to weld U plates onto as they rarely suffer any significant attack and, therefore, provide long and continuous strong points.

Plating plus reinforced concrete infill – used for repairs to side and in-pans, usually on Z piles. Typically, steel reinforcing hoop bars are welded to the piles. Similar bars are also welded to the back of a plate which is then welded across the outpans to span the inpan (see Figure 6.2). Additional reinforcement may be placed in the void prior to placing the plate. Once the bottom is sealed, concrete is poured to fill the void behind the plate. This technique is usually unsuitable for U pile repairs. This is because the repair plate is required to be welded to the centre of the outpan which is often the location of ALWC attack for this pile geometry. For Z piles the weld is to the clutch which is unaffected by ALWC for this pile geometry.

Box 6.6 ***Case study for welded strengthening plate on solid structure (Steele, 2002)***

A typical project involving this type of repair method on a SSP wall at the Port of Southampton comprised the following procedure:

- *use of limpet dams to inspect and repair U-shaped sheet pile wall structure under dry conditions*
- *following removal of marine growth, ultrasonic thickness readings were taken to identify which piles required to be repaired. The results revealed that up to 8.5 mm steel (equivalent to 0.9 mm/yr) had been lost from front face and occasionally corners of outpans in a zone 1.0 m high centred at 0.5 mCD. Elsewhere on piles in the submerged zone virtually no corrosion was detected.*

The piles were first cleaned by grinding over the repair area following which 2050 mm long and 22 mm thick mild steel plates were profiled to a U-shape to fit over the outpan and welded back to the outpan clutch of the adjacent piles. It was reasoned that 10 mm would replace material already lost and 12 mm would act as future corrosion allowance which might notionally be worth 13 years.

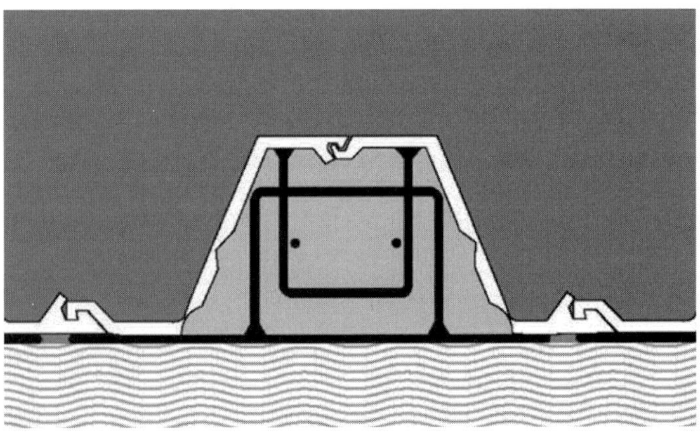

Figure 6.2 **Plating and concrete infill repair for Z pile**

Box 6.7	Case study for plating plus reinforced concrete infill on Z piled wall (Steele, 2002)

An example of a typical project involving this repair method has been used at the Port of Felixstowe and comprised the following procedure:

- *limpet dam suspended from a 60 ton crawler crane with capability of spanning three pairs of Z piles allowing actual work on two piles at a time*
- *thickness testing and identification of piles to be repaired was carried out following removal of marine growth. Given the overall limited life of the quay wall, areas having holes or thicknesses below 6 mm (50 per cent) on sidepans or 8 mm (50 per cent) on in/out pans were designated for repair*
- *all holes were initially plugged by high density cellular foam following which a steel plate was fixed to the wall via welded screw studs*
- *following sealing of the holes, 25 mm diameter reinforcing rods were welded to the sidepans and a 12 mm plate with a bottom plate soffit and 25 mm bar welded to the back were fixed to the reinforcement and welded across the outer pans. The resultant void was infilled with cement grout to provide additional bending strength*
- *cleaning, repair of holes and thickness inspection and resultant repairs were carried out in one operation before relocation of the limpet dam.*

Reinforced concrete facing – structural repair comprising an in-situ concrete facing. Steel reinforcing bars are welded to the piles and used to support the wall reinforcement. The face is shuttered and concrete poured to form a partial or continuous wall. The shuttering is then removed.

Box 6.8	Case study for reinforced concrete facing (Sandison, 2001)

An example of a typical project involving this repair method was used at Lerwick Harbour in Shetland, Scotland although, in this case, the damage was not only as a result of ALWC. Extensive holing and pitting of the steel inpan surfaces from the splash zone down to the mud zone was reported over a 1.4 km length of sheet piles. Shuttering was first placed and the concrete mix was pumped a height of 13 m from the seabed to infill the inpans. A corrosion inhibitor was added to the concrete mix since there was concern that the surface of the steel wall and reinforcement would be heavily contaminated with salt. Polypropylene fibres were also added to the concrete mix designed to minimise shrinkage cracking.

Pile splicing – while technically feasible, the removal of a length of steel sheet piling and the splicing in of a replacement section at low water level and above presents considerable practical difficulties. In particular, it will be necessary to remove (or temporarily support) the fill material while the replacement is undertaken and then replace it on completion. The corrosion implications are similar to those for repair plates (see Section 6.4).

Replacement sheet piling – where piles are beyond economical repair and/or the structure is unsafe and where the existing structure permits, a row of shorter piles can be driven in front of the affected piles and tied into the existing structure, alleviating the need for total reconstruction. Accelerated corrosion attack is possible on the new section of piling if it is electrically bonded to the old structure through the tie back fixings. However, the gap between the rows of piles is normally filled with concrete which acts as a protective barrier in the event of holing.

If the new piled wall is limited to just covering the low water zone then this could be a hazard for shipping unless suitable offset fendering can be introduced without affecting the reach of the cargo handling equipment. The method is particularly useful for sea defence works.

Box 6.9	**Case studies for replacement sheet piling**
This technique has been implemented at the Ports of Felixstowe (Steele, 2002) and Southampton (Holleywood, 1978). In the latter case a row of sheet piles was constructed in front of old masonry and concrete walls and the resulting void filled with concrete. Other examples have also been reported (in Germany- Reinhard, 1998) and Australia where an impressed current CP system was installed to the rear of the new piles to prevent a macro cell between the old and new structures (see Appendix A11, ALWC Case Study 5).	

6.3 Access to carry out works

A safety risk assessment should be a key part of the choice between access methods (see Sections 4.7 and 5.2.4). The quality of the work will be highly dependent on the ease and security of access. To a large extent, this will depend on the complexity of the structure and the consideration given by the designer to future safe access to undertake inspection and maintenance works (see Appendix 9). Efficient access for carrying out in-situ repairs (and installation of protection systems) can be made difficult, if not impossible, by inconsiderate initial design. Safe and efficient access to carry out works is a CDM requirement that designers must consider during the development of their designs, the hazards and risks which may arise to those constructing, cleaning and maintaining the structure.

The repair work will have to be undertaken at low level and will be severely restricted by the tide. The available options are therefore:

- underwater repairs undertaken by divers
- dry repairs undertaken within a temporary dry environment formed using a limpet dam or cofferdams.

The advantages and disadvantages of these two methods are summarised below. Further information on these and other access methods are given in Appendix 7.

Table 6.2 **Comparison of underwater versus dry repair methods**

Underwater repairs	Dry repairs
Access to affected areas at most states of the tide	Access to the affected areas at any state of tide
Quantification of damage less accurate than in the dry	Accurate quantification of corrosion damage
Work can be prevented by poor underwater visibility	Work not affected by poor underwater visibility
Work can be prevented by currents or waves	Work not affected by currents
Quality of work can only be verified by a qualified diver/inspector	Quality of work can be verified by owner's or engineer's usual qualified inspectors
Quality limited to underwater standards	Better quality of repairs
Flexible location of working	Location limited to a discrete length of quay at any one time

No additional loading on the structure	Requires the structure to be able to support the uninstalled weight of the cofferdam, operational loadings and of the crane needed to lift it into place
Safety zone required between port operations and divers	Minimal disruption to port operations due to the flexibility of movement (Steele, 2002)
Low mobilisation costs	High mobilisation costs – not cost effective for small repairs.
Not able to recover any possibly contaminated washings from going into seawater	Any contaminted washings can be collected

6.3.1 Underwater repairs

It is important to arrange a suitable, secure position from which the welder/diver can work. Permits must be gained from the regional environmental authority for any underwater work.

In the case of steel sheet piled walls this can be achieved using a small mobile crane to lower the diver in a man rider basket down the face of the wall to the required depth. Additional restraint will often be required when undertaking operations such as high pressure water jetting, grit blasting or grinding. Actions can be taken to counteract strong reactive forces upon a diver. Often supported platforms (similar to window cleaning gantries) are also used.

For skeletal structures, this can be achieved using a de-mountable work platform, for example, two semi-circular buoyancy tanks hinged together and secured around a pile. Flooding the tanks allows the platform to sink to the required level and purging the tanks of water allows it to be re-floated.

In principal, repair methods used above water can also be used underwater by an experienced diver using specialised tools and techniques. However, the repairs are more time consuming and difficult to complete and inspect than those performed in the dry.

The initial cleaning of the structure within the low water zone can be carried out using normal high pressure water jetting equipment to a depth before water pressure starts to reduce its efficiency. However, this is unlikely to be a problem when using the extra high pressure equipment now available.

Welding underwater can be achieved by an experienced welder/diver using special flux covered, waterproof electrodes. There is a limited choice of welding materials available for underwater welds. While the more expensive, but better quality, underwater electrodes can give superior results to normal electrodes, underwater welds will always be inferior to welding in dry conditions. In particular, structural welding will be impossible underwater if visibility is poor and the welder/diver has difficulty seeing the work area. With wet welding it is impracticable to bridge a gap of more than 2 mm and, therefore, it is not unusual for divers to use steel strip ground to shape to fit into the gap before welding. Although underwater welding can return some structural strength to a weakened element and it is cheaper than limpet dams for small volume of works, it should only be considered for non-structural welding jobs such as for emergency patch repairs and the attachment of galvanic anodes to a structure. A typical case study involving underwater welding of repair plates using divers is given in Appendix 11, case study 2.

6.3.2 Dry repairs

Inspection and repair below the waterline can be undertaken in the dry using limpet dams (see Plate 6.3). These structures which effectively provide a dry environment along a discrete length of a steel sheet piled wall, or on an individual pile of a skeletal structure, are discussed in detail in Appendix A7. Potential advantages using this form of access include:

- access to the damaged zone at any state of tide
- more accurate quantification of corrosion damage
- better quality repairs
- verifiable quality of the repair
- access for the owner and engineers to assess the damage and subsequent repairs for themselves.

See Plate 7 *Limpet dams in place on steel sheet piled wall* on page 101.

6.4 Corrosion implications of repairs

Many types of repairs such as the steel plate welding method or partial jacketing with reinforced concrete can introduce new corrosion cells on the structure and, in some circumstances, locally increase the corrosion rates. These can be effectively suppressed by the installation of a suitably designed partial or complete CP system (see Appendix 10) although sometimes alternative, more cost-effective, methods may be available in specific circumstances. An example of the latter would be the application of a suitable coating to cathodic (non-corroding) surfaces in the tidal zone on a sheet piled wall to reduce corrosion on surfaces in the low water zone (see Appendix 2, Section 2).

In the case of fixing repair plates, increased corrosion on the repair materials used may occur due to crevice corrosion (eg due to incomplete welds), bi-metallic corrosion (new versus old steel), PWC on weld components and, to some extent, cold bending of the repair plate material to produce the required profile; the higher stressed bend areas can become anodic (ie corrode more) than the surrounding less stressed areas.

Measures to minimise the risk of crevice attack have already been discussed in Section 6.2.1 and involve filling any voids between the edge of the repair plate and the substrate steel with non-shrink cement grout or an epoxy based paste. It should be noted, however, that there is little hard evidence that 'classical' crevice corrosion (such as that which occurs with stainless steel in seawater for example) will occur with carbon steels. Consequently the precautions suggested, such as the sealing of crevices or the protection of unsealed crevices may not be necessary. This is an area that needs further research (see Chapter 9).

The important factors that will influence bimetallic corrosion in seawater are the potential difference between the old versus new metal (ie the driving voltage) and the area relationship between the metals (which will vary with tides). The repair plate will be the component on which the increased corrosion will normally occur and, as a result of the unfavourable area ratio between it and the surrounding structure, the corrosion rate will be aggravated. The following options are recommended for repair plating:

1. The use of standard mild steel but as thick as can be practically fitted or bent to allow it to suffer increased corrosion without loss of efficiency. Standard mild steel

plate (ie complying with EN 10248) is readily available, cheaper and often easier to weld and bend than the low alloy steel grades. However, if heavier sections are used, they can be less manageable and have safety implications in installation. In such circumstances, applying a coating to the repaired area on an uncoated structure would not be recommended since it may well result in rapid pitting type attack at "holidays" or damaged areas in the coating.

2. The use of standard mild steel but the structure protected by the local application of a coating, covering and extending beyond the repair, in combination with local CP. Local fixing of a limited number of galvanic anodes adjacent to the repair would be expected to extend the life of the repair and the welds. The coating will reduce the rate of consumption of the anodes which otherwise would probably be rapid. More regular monitoring and timely replacement of anodes will be required for this option (see Chapter 7).
3. The use of standard mild steel and incorporating a full cathodic protection system installation into the repair works.

Welds and fasteners (such as fixing bolts) are critical components and should not be allowed to corrode preferentially ie they should by designed to be cathodic to the surrounding steel structure. Preferential weld corrosion (PWC) is a form of bi-metallic corrosion which can occur either on the weld bead or in the adjacent heat affected zone (HAZ). The risk of such attack can be minimised to a great extent by the correct choice of welding consumables. Also maximum hardness values for welds should be specified where impressed current CP systems are to be used in order to minimise the risk of hydrogen embrittlement in the event of overprotection (see Chapter 7, Section 3.1). Fixing bolts are usually fabricated using high strength, low alloy steels which would be expected to be slightly cathodic to standard mild steels and, therefore, should be protected.

Some caution is also necessary in respect of steel encased in concrete contacting with steel in sea water, particularly if some or all of the steel in concrete is in the tidal zone (see Appendix 10, Section 3).

Note that adding plates or concrete encasement to counter the effects of corrosion can create localised stiffness on a sheet piled structure. If the initial design was carried out on the assumption of a particular wall strength and stiffness, changing that stiffness could invalidate this design assumption. This could become more problematic in applications where walls are designed plastically.

6.5 Protection options for new build and existing structures

This section considers the measures that can be taken to:

- minimise the risk of ALWC occurring on new-build or replacement structures
- treat existing structures in the early to intermediate stage of ALWC attack or in an advanced stage after repair.

The corrosion protection strategies that are applicable to ALWC are those based on well established conventional methods, primarily CP, coatings of various types and concrete cladding or a combination thereof (most commonly CP in combination with coatings). More detailed information on these protection methods is provided in Appendix 10. These latter techniques have a long track record for protecting steelwork against the common forms of corrosion in the marine industry and all can be expected to contribute to the protection against ALWC to a greater or lesser degree. All have their own proponents and those with commercial interest in promoting them. There are surprisingly few remedial techniques aimed at treating ALWC in a more focused rather than general way, although one is currently under development and is briefly considered in Chapter 9, Section 9.2.

The CIRIA survey carried out as part of this book (see Chapter 3) indicated that the most popular methods used to protect against ALWC, although not always successfully, were reported as follows:

Protection method	% use
Galvanic CP	38
Coatings	30
Welded patch plates	15
Impressed current CP	8.5
Concrete	8.5

Since ALWC is a recently discovered phenomenon, some involved in finding an immediate solution to prevent or treat ALWC on particular structures have, by necessity and often in isolation, developed a range of corrosion control strategies using a myriad of commercially available methods and products. Due to the recent nature of the problem, specific information on the long-term effectiveness of such strategies is largely unavailable. It is, therefore, important that independent professional advice is sought to give guidance on the best way forward for any particular structure affected by ALWC. It could reasonably be expected that this will require a broad spectrum of engineering expertise with a structural engineer relying on other specialists including corrosion engineers, coating and/or CP specialists, metallurgists etc in order to obtain the most appropriate solution.

The selection of the most suitable corrosion control strategy for dealing with or reducing the risk of ALWC (ie the lowest cost that meets the particular requirements) will be different for each structure and location. The particular circumstances should be evaluated on an individual basis taking into account the merits and risks of each potential strategy. Only general guidance on protection strategies can be given here. The final selection will depend on various factors including:

- *risk* – of ALWC occurring or reoccurring for new build or replacement structures
- *extent and distribution* – of ALWC for existing, already affected, structures
- *economics* – financial resources available for maintenance and corrosion control; including application and maintenance (ie life cycle costs)
- *effective life* – how long it performs its protective function compared with the design, or remaining, life of the structure
- *accessibility* – whether the protective measure can be satisfactorily applied, inspected, monitored and maintained
- *disruption* – whether the protective measures can be implemented with minimal disruption to operation from installation
- *extent of protection required* – given that the effects of ALWC are highly localised, the options are complete (ie applied to the whole structure), partial (ie applied in specific zone or zones) or focused protection (ie local at ALWC patches)
- *personnel* – availability of on-site, trained personnel for regular inspection, maintenance and monitoring duties
- *health and safety issues*
- *environmental implications.*

6.5.1 Protection methods

The principal conventional methods available for protection of new build and existing steel structures are as follows:

1. **Corrosion allowance**

- specifying a thicker/heavier steel section than is required to achieve structural requirements.

New build or replacement structures	Existing structures	Repair works
• makes assumption on the corrosion rate to determine the thickness of additional steel required. Where long life structures are required, this would probably necessitate an impractically thick section to deal with high corrosion rates associated with ALWC. Therefore, use in conjunction with other protection method such as CP or coatings or CP in combination with coatings.	N/A	• generally use the maximum steel thickness for repair plates determined by ability to bend the plate to the required profile.

2. **Steel grade**

- use the allowable stresses appropriate to mild steel (eg S270GP) but adopt a higher strength steel (eg S320GP or S355GP)
- high strength steel is unlikely to provide any additional protection against ALWC compared with mild steel.

New build or replacement structures	Existing structures	Repair works
• provides a greater factor of safety against structural failure and increases the effective working life of a given section • the formation of holes with possible loss of fill would probably not be delayed • deflection of the corroded section also requires to be considered.	N/A	• advantageous to specify low alloy steel composition which is more electrochemically noble than surrounding old steel. Alternatively use thicker section of mild steel (see above) • possible increased difficulty in bending and/or welding low alloy steels

3. **Welded patch plates**

- increase the section thickness in the low water zone, by attachment of profiled steel plates (plates typically 2-3 m in length welded to the outpans of "U" piles or across the front of the inpans of Z pairs)

- given the potentially rapid rate of loss of section associated with ALWC (and bimetallic corrosion), plating should be used in conjunction with other protection methods such as CP or coatings or CP in combination with coatings (see Section 6.4) to provide longer term protection
- attention needs to be given to the choice of welding consumables and procedures to minimise the risk of PWC.

New build or replacement structures	Existing structures	Repair works
• makes assumptions on where the ALWC will occur and the corrosion rate. There is evidence that ALWC may occur below LAT • steel plates are more easily and cost effectively welded to the pile prior to installation • can cause problems when driving piles • relies on the steel sheet pile being driven to preset level. If the pile hits an obstruction, the patch plate may end up at the wrong level • see ALWC case study 6 (Appendix 11) for practical example.	• steel plates can be fitted to piles at any time • used to provide additional thickness at thinned areas • see ALWC Case studies 1 and 2 (Appendix 11) for practical examples.	• used for non-structural repairs making good holing to prevent or halt the loss of backfill and/or the ingress of seawater • it is not suitable for more extensive areas of damage where excessive thinning and/or holing has been identified over several pile widths with limited sound metal to weld to.

4. Structural modifications

- delay the loss of structural integrity by moving the location of maximum bending stress away from the critical low water zone area.

New build or replacement structures	Existing structures	Repair works
• could be achieved, for example, by adjusting the level of walings and tie rod system. There is evidence that ALWC may occur below LAT • holing and loss of infill will probably still occur but the structural significance will be reduced.	• could theoretically be considered, but unlikely to provide an economical solution.	N/A

5. Cathodic protection:

- both galvanic anode and impressed current CP systems have established and successful track records for protecting immersed steelwork against marine corrosion, including MIC

- if properly designed, installed, commissioned and maintained, it can be fully effective against any form of corrosion up to about the mid tide zone
- ALWC has not been noted on any structure where a well designed and fully maintained CP system had been installed at the construction stage.
- there is currently no design guidance on current density requirements for protecting existing structures affected by ALWC. However, designs at the upper range of existing guidance appear to provide adequate levels of measured protection potentials and should, therefore, be used until further clarification is obtained by research and/or experience (see Appendix 10, Section 2.2b)
- regular inspection and maintenance is required (see Chapter 8) more so with impressed current systems than galvanic anode systems
- can be used either alone or in combination with other protection methods (eg coatings)
- impressed current CP systems normally needed to be used (1) in the potentiostatic (ie constant potential) rather than the more common galvanostatic (ie constant current) mode or (2) in combination with coatings (applied in the tidal zone), to minimise the effect of changes in surface area to be protected due to tides. On deep berths with low tidal ranges, this may be less significant
- the use of zinc galvanic anode systems may not be appropriate in certain situations where a facility is more confined and where zinc levels in the water are already close to maximum permitted levels (Matthiessen, 1999 and Bird, 1995). Aluminium galvanic anodes are available with significantly reduced zinc levels (eg Al-Zn-In alloy anodes typically have a four per cent Zn content)
- Al-Zn-In galvanic anodes require a certain level of chloride to activate and may not operate effectively when used in brackish waters.

New build or replacement structures	Existing structures	Repair works
• either galvanic CP or an impressed current CP system can be used in most situations (see table of CP Case studies Appendix 11 for practical examples). The choice is often determined by (1) the design life (2) the mechanical robustness required and (3) the availability of resources, both funding for the initial installation and suitably trained personnel for the long term monitoring and maintenance, particularly in the case of an impressed current CP system • CP can either be installed or planned for when the structure is built. The latter approach involves preliminary engineering by a CP	• either galvanic CP or an impressed current CP system can be retrofitted to an existing structure • it is recommended that a relatively long life for the CP system is proposed because installation is a large proportion of the cost • can be used in combination (ie a hybrid system), for example, to help solve "geometry" difficulties where the distribution of CP current is difficult • special attention is needed where an existing deteriorated coating is present on the structure (ie possible current shielding effects below disbonded coatings) • the current density required to arrest	• the provision of partial/localised CP by local fixing of a limited number of galvanic anodes adjacent to the repair to extend the life of the repair and the welds. Applying a suitable coating will reduce the rate of consumption of the anodes which would otherwise be rapid. More regular monitoring and timely replacement of anodes will be required for this option (Chapter 9).

specialist and construction related preparations for subsequent installation (see Appendix 10)	ALWC on existing structures and to maintain that condition is likely to be greater than that required for general corrosion considered in most CP standards (King, 1989) • see ALWC case studies 2 and 3 and table of CP case studies, Appendix 11, for practical examples.	

6. **Paint coatings**

- have track record for the protection of maritime, steel structures continuously or intermittently immersed in seawater, while coating remains intact
- coating inspection during application and regular inspection and maintenance during service is required (see Chapter 7)
- coating application options are:

  (1) to apply over the whole exposed surface of the structure down to the low water zone.

  (2) as (1) but to some distance below the low water zone using a limpet dam.

  (3) limit to the low water zone only using coatings that can be applied underwater.

- less suitable where surfaces exposed to mechanical damage or abrasion
- risk of damage during installation
- if used in combination with CP, it is vital that the coating should be compatible (see Appendix 10, Section 4)
- coatings should have good resistance to biodegradation, chemical and microbial attack
- although still recognised in the latest standards in the UK (eg ISO 12944), coal tar products are known to be a potential health risk and should not be specified where more acceptable alternatives are available

New build or replacement structures	Existing structures	Repair works
• option (1) recommended in conjunction with CP • the coating can be field applied (ie at the time of the construction or after driving of the pile) or applied under factory conditions (ie before driving the pile) with on-site repairs to any damaged areas • factory applied coating systems should be VOC compliant meeting the current requirements of PG6/23.	• in the absence of CP, specifications normally require Option (2), typically from the top of the pile to 1m below LAT • steel surface condition in the submerged zone should be checked to confirm the required depth of coating for each application • options (1) and (2) can be used in conjunction with CP if the coating is compatible, applied correctly and is well bonded	• option (3) problematic due to high risk of pitting attack if coating applied to the repair area only • option (3) can be used in conjunction with local galvanic CP anodes (see above). The main purpose of the coating will be to reduce the rate of anode consumption • coatings are available that can be applied and cured underwater but effectiveness can be variable. Since 1995

• field applied coatings will always be inferior to factory application.	• field applied coatings will be inferior to factory application due to the inevitable difficulties that will be encountered in surface preparation • for high performance coatings, surface preparation and application should be performed under dry conditions using a limpet dam • where blast cleaning is impossible more surface tolerant coatings are available which may be applied over under prepared surfaces using less rigorous methods of cleaning (eg hand tool, power tool cleaning etc). The durability of such coatings will be less than for high performance coating systems • see Appendix 11, case study 1 for practical example.	reformulation of some epoxies for underwater application (ie specifically to overcome health risk problems associated with specific curing agent) has reportedly resulted in significant deterioration in ease of application and performance properties (Semerad, 2003).

7. **Wrappings/tape and encapsulation protection systems**

New build or replacement structures	Existing structures	Repair works
• may be considered, but not usual.	• well established for application to skeletal structures particularly in the splash zone but can be installed both above and below water • surface tolerant systems • difficult to inspect steel surface and efficacy of protection after application.	• applicability on repaired tubular piles.

8. **Metal spray coatings**

- usually aluminium or zinc coatings with suitable sealer topcoats to penetrate and seal the porosity. A paint coating system can also be applied over the metal spray to provide further protection
- aluminium rather than zinc metal spray coatings are preferable from both a performance and environmental perspective
- factory applied metal spray coatings with good quality control have a track record for the protection of offshore steel structures continuously or intermittently immersed in seawater.

New build or replacement structures	Existing structures	Repair works
• can give effective long term protection if applied to the whole surface before installation • can contain high porosity which is not always possible to seal adequately • may suffer accelerated attack in service if surface only partially coated or if used with improperly controlled impressed current CP • can significantly reduce current required for galvanic CP.	• cannot be readily applied to an installed sheet piled wall.	• could be applied but would suffer accelerated attack (as for galvanic anodes) if restricted to repair area.

9. **Concrete jacketing/encasement:**

- has track record for protecting solid and skeletal structures with limited movement, mechanical damage, etc to minimise risk of concrete cracking
- good quality concrete required with adequate cover (typically 50 mm to 100 mm) and appropriate additives
- pre-coated steel reinforcement can be used but poor track record of efficacy of long term corrosion control for badly designed and installed systems
- both precast concrete elements and underwater concrete (cast in-situ using divers to install and remove shuttering) have been used
- the number of joints should be minimised and special attention should be given to the seals at construction joints
- possibility of bacterial attack on the concrete itself
- beware of introducing corrosion concentration cells if only partial encasement/jacketing is applied (see section 6.4 and Appendix 10, Section 3). To prevent, either use CP to protect the steel below the jacket or apply a suitable coating to the steel reinforcement and the part of the steel pile wall surface within the concrete.

New build or replacement structures	Existing structures	Repair works
• has mainly been used in the past for partial protection in the atmospheric, splash and tidal zones (with galvanic CP) • can be used for full protection with the concrete extending down to bed level, in order to minimise the risk of introducing additional corrosion cells but this may not be cost effective.	• can be used to partially protect existing structures in conjunction with galvanic CP • concrete jackets to bed level have also been successfully applied on tubular piles • protection in the low water zone can be achieved by welding plates over every outpan (or a number of outpans) and filling	• welding of plates and concrete infill specifically at thinned and/or holed locations.

	the void behind with concrete • as above but without the plates. Typically, steel reinforcing bars are first welded to the inpans of the piles, the face is shuttered, the concrete poured and allowed to cure and finally the shuttering is removed.	

6.5.2 Protection strategies for new build structures

Historically, the design of new build, steel structures for the maritime industry has relied on thicker steel sections to give the required design life. The expected life would routinely be predicted from corrosion rates given in design standards such as BS 6349 or from steel manufacturer's technical literature. However, as a consequence of the rapid deterioration effect of ALWC, such design standards and guidance documents now quote potential corrosion rates in the low water zone an order of magnitude higher than previously reported. Trying to deal with these high, localised corrosion rates using a simple corrosion allowance/additional steel thickness is rarely practical where long life structures are required. For most situations this makes some additional form of corrosion protection necessary in order to achieve the required design life which is typically between 25 to 100 years. Since ALWC occurs randomly, it will first be necessary to make an informed decision on the risk of ALWC occurring at the particular site and how any occurrence will be managed during the life of the structure. Unfortunately, no reliable predictive model for assessing the probability of ALWC occurring is currently available (see Appendix 4) and because of the apparent random nature of the problem between sites, knowledge of the local incidence of ALWC, while helpful, cannot be relied on. The options available to the designer/engineer for new build structures are therefore:

1. To install protection measures at the construction stage, on the basis that ALWC will occur and the cost of this initial protection will be less than repair works at a later stage – the *"prevention is better than cure"* approach. Not being able to confidently estimate the risk of ALWC occurring on the structure may make it more difficult to justify the expense of such preventative measures. However, such protection measures are likely to be applied across large parts of the structure and, therefore, this approach should provide protection against other forms of corrosion in addition to ALWC.
2. Design contingency measures into the structure to make it easier and quicker to inspect, repair and implement suitable protection in a cost-effective manner at a later date (if necessary) and implement an ongoing monitoring regime – the *"watch and see"* approach. Even if ALWC does occur then it is likely that there will be an induction period during which time corrosion will be less aggressive and initial steel loss less critical. Also other types of remedial measures may become commercially available in the future to treat ALWC in a more highly focused and cost effective way than is currently available (see Chapter 9).
3. Ignore the possibility of ALWC occurring – the *"do nothing"* approach. The apparent increasing spread of ALWC, its potential rapid deteriorating effects and the high cost of repairing or replacing an affected structure, rarely makes this a serious option.

Design measures to consider on new build structures include:

- electrically bonding individual elements such as piles. Where a concrete capping is used, bonding should be done before casting the cap
- installing ancillary equipment (eg brackets for galvanic or impressed current anodes and reference electrodes as required), establishing routes for cable shafts and ducts and the positions for protective current supply units, distribution boxes and monitoring units etc (for impressed current systems only) to facilitate the installation of CP at a later stage, if required
- use CP anode fixings that will allow inspection and replacement without the need for a diving team
- keep the design details as simple as possible
- avoid areas which are difficult to access (eg pile clusters where the space between individual piles is restrictive) to facilitate inspection and maintenance
- consider the use of U piles rather than Z piles (the former will intrinsically last longer due to their heavier section thickness) where appropriate. However, Z piles are now available with thickened corners (where ALWC effects are most concentrated)
- installing corrosion monitoring systems (see Chapter 7).

6.5.3 Protection strategies for existing structures

The options available to the engineer for existing structures affected by ALWC are:

1. Carry out repair works to patch all holes and reinforce all thinned and weakened areas followed by installation of suitable protection measures to minimise any future deterioration – the *"preventive protection"* approach. This would require utilising the results and recommendations of the detailed survey report to regain structural adequacy (see Chapter 5) followed by implementation and ongoing monitoring/maintenance of the most appropriate protection system (see Section 6.5). If the protection can be applied in a focused, rather than in a general way, the costs and maintenance burden can be significantly reduced (Bailey, 2004)
2. Perform minimum essential remedial works to maintain structural stability – the *"monitoring and repair in time"* approach. This would require a regular, on-going inspection/survey and repair programme patching holes and strengthening critical thinned areas, as necessary. While such works will undoubtedly increase the life of the structure, ALWC would be expected to continue unabated, requiring frequent and costly repair works.
3. Allow the structure to gradually deteriorate until it becomes unserviceable – the *"do nothing"* approach. This is rarely a serious option but where it is applicable to a particular structure then regular monitoring is necessary as an absolute minimum to ensure structural integrity remains within the appropriate factors of safety.

SUMMARY

- a typical logic sequence for the selection of repair strategy for both skeletal and solid steel maritime structures is given in Figure 6.1
- repair techniques available for both skeletal and solid structures include welded patch repair or strengthening plates, systems utilising concrete, splicing and replacement piling
- efficient and safe access is required to facilitate in-situ repairs and installation of protection systems
- underwater welds should only be considered for non-structural welding applications dependent on welder ability and experience
- most types of repairs can introduce new corrosion cells onto the structure and appropriate measures should be taken, as necessary, to minimise this risk
- the most suitable protection strategy will be different for each structure and location. It is important that independent professional advice is sought to give specific guidance on the most appropriate strategy for any particular structure
- general guidance on protection strategy depending on the condition of a structure is tabulated below:

Condition of structure	General guidance
New build or replacement structures	A minimum risk, "standard" approach for corrosion protection (including against ALWC) would typically be the application of a suitable high performance coating system to the atmospheric, splash and tidal zone areas in combination with a well designed, installed and commissioned CP system, probably using galvanic anodes for ease of use, reliability and low maintenance (see Appendix 10). Ongoing monitoring and maintenance of the corrosion protection system will be necessary throughout the life of the structure but this will be reduced with a galvanic anode system. (see Chapter 7). Alternative approaches to design and protection, often using measures outlined in Section 6.5, can also be considered depending on particular circumstances. The need for more regular inspection and monitoring will increase as the overall level of corrosion protection is reduced.
Existing structure – early to intermediate stage	A prudent approach would be to initiate a regular inspection programme to facilitate planning; preparation and budgeting for subsequent corrosion protection (see Chapter 5). Steel renewals will be necessary, particularly at critical areas, where significant metal wastage has already occurred (see Sections 6.1 and 6.2). Seek independent, professional advice for guidance on the most appropriate and cost effective corrosion protection strategy and when to apply it. The aim should be to prevent the structure from reaching an advanced stage of deterioration.
Existing structure – advanced stage	A prudent approach would be to initiate urgent repairs, or modified use, as described in Sections 6.1 and 6.2, followed by protection against future corrosion. Seek independent, professional advice to give guidance on the most appropriate and cost effective corrosion protection strategy for the particular structure.

Plate 1 ALWC holing at low water on typical maritime structure

Plate 2 Close up of typical ALWC patch

Plate 3 Holed "Z" section pile showing loss of backfill material

Plate 4 Close-up inspection using platform and crane

Plate 5 Thickness measurements being taken on steel sheet piled wall using a digital ultrasonic thickness meter

Plate 6 Repair using welded "U" shaped steel plates on steel sheet piled wall

Plate 7 Limpet dams in place on steel sheet piled wall

7 Monitoring of performance of preventative, remedial and strengthening measures

This chapter will review the methods available to monitor the performance of the candidate remedial systems referred to in Chapter 6 for dealing with structures affected by ALWC. Accurate monitoring is particularly important since many of the remedial methods outlined in Chapter 6 for dealing with ALWC are not established enough to have proven, long term track records.

The monitoring methods proposed are for guidance purposes only. Operating and maintenance procedures provided by the installers/suppliers of the remedial method should be followed to ensure operational efficiency of the installed systems.

The records of inspections, examinations and repairs should be accurately completed and preserved for use in future inspections/surveys as part of the asset management method (see Chapter 4).

7.1 Monitoring methods

The methods used for monitoring repairs and protection systems are the same as those described in Chapter 5 and Appendix 7 and include:

- visual and tactile
- electrochemical surveys
- photography, video, etc
- physical measurements (eg micrometers, pit depth gauges, etc)
- ultrasonic thickness measurements and flaw detection
- on-site sampling and analysis.

7.2 Post implementation assessment and evaluation

1. Repair plates and welds

An inspection of these areas should be carried out as part of a continuing monitoring programme for the structure.

- Visual evidence for:
  1. Any re-occurrence of ALWC (ie iron sulfide rich surface deposits, shiny, pitted substrate).
  2. Bi-metallic attack (ie more uniform metal wastage across total surface area) on uncoated repair plates.
  3. PWC on welds (ie preferential loss of weld metal or evidence of grooving in HAZ or parent metal adjacent to the weld).
  4. Cracking at welds, particularly if impressed current CP is in operation (see Section 3.1 below).
  5. Pitting attack at defects in any coating applied to repair plates (indicative of galvanic attack).

- ultrasonic thickness and/or pit depth gauge measurements for monitoring ongoing corrosion rate in order to check that the actual rate is less than planned rate and/or to establish remaining life.

2. Coatings

- an inspection of any coatings applied to the structure and coatings associated with any CP installation (eg impressed current anode dielectric shields or support tubes) should be carried out as part of the continuing monitoring programme for the structure in order to establish the extent of necessary coating repairs. If a CP system is fitted it may not be necessary to repair all, if any, of the coating damage. The aim of the monitoring is a general assessment of the protective coating system by looking for signs of:
  1. Rust staining (orange discolouration).
  2. Mechanical damage.
  3. Coating disbondment.
  4. Underfilm corrosion particularly around damaged coating areas and exposed edges where coatings have only been applied to parts of the structure.
  5. Blistering.

- the degree of coating degradation in the absence of CP can be assessed in accordance with ASTM D610 (ASTM, 1968)
- all repairs to coatings should be carried out in accordance with the coating manufacturer's requirements.

Table 7.1 **Explanation of ASTM D 610 Ratings (reprinted, with permission, copyright ASTM International)**

Rust grades	Description of rating
10	No rusting or less than 0.01% of surface rusted
9	Minute rusting less than 0.03% of surface rusted
8	Few isolated rust spots less than 0.01% of surface rusted
7	Less than 0.3% of surface rusted
6	Extensive rust spots, but less than 1% of surface rusted
5	Rusting to the extent of 3% of surface rusted
4	Rusting to the extent of 10% of surface rusted
3	Approximately 1/6 of surface rusted
2	Approximately 1/3 of surface rusted
1	Approximately ½ of surface rusted
0	Approximately 100% of surface rusted

Table 7.2 **Approximate correlation between ASTM D 610 and ISO 4628/3 (BS 3900:Part H3:1983) (reprinted, with permission, copyright ASTM International)**

ASTM D 610 Rust Grade	ISO Rust Scale
10	Ri 0
9	Ri 1
7	Ri 2
6	Ri 3
4	Ri 4
1 to 2	Ri 5

3. Cathodic protection

The inspection and control requirements of galvanic anode and impressed current cathodic protection systems are different (see also Appendix 10, Section 2).

Galvanic anode CP systems, if properly designed and installed, will "self control" their current output to maintain a relatively steady level of cathodic polarisation (protection level) to the steel which should be more negative than -0.900 Volts (with respect to a Ag/AgCl/seawater reference electrode) across the structure, measured without IR errors (or "Instant OFF" potential). This "self control" will accommodate high and low tides, variations in flow rates, storms and inter-connected vessels if the anodes have been ideally sized in accordance with the present design codes (principally EN 13174: 2001). If the anodes are securely welded to the structure, their alloy composition is within appropriately specified limits and they do not suffer mechanical damage, the system is generally maintenance free until it approaches the end of its design life and the anode consumption dictates that the anodes are too small to meet their peak current demands. There are no control or adjustment requirements.

It is, therefore, established good practice with galvanic anode CP systems to verify their performance at commissioning (often one month following completion of installation) and possibly again at the completion of the maintenance/defects liability period (typically 12 or 24 months). These measurements will often include steel/seawater potential measurements at different depths on many or all elements within a structure (ie proving the distribution of adequate cathodic polarisation and the electrical continuity between elements of the structure) and may include "Instant OFF" measurements on coupons to ascertain the extent of IR errors due to cathodic protection current flow in the measurement circuit (although these are normally insignificant with galvanic anode systems). Limited representative measurements may be taken at three monthly intervals to determine the progress of time dependent polarisation and the effects of sea state and temperature variations. Thereafter performance verification may repeat some or all of these measurements at intervals of one to three years in order to ensure that any variations in the performance are detected at an early stage. When the anodes are reaching the end of their design life it is normal for selected representative anodes to be cleaned and inspected by a diver so that an estimate of residual mass and life of the anodes can be made.

Impressed current CP systems rely on the availability of continuous ac power to the transformer-rectifier. To accommodate high and low tides (particularly for structures with high tidal ranges) they are designed and set to operate "automatically" in a potentiostatic (ie constant potential) mode and may depend on the reliability of the permanent reference electrodes and their associated monitoring cables for continued operation. It is established good practice with impressed current CP systems to undertake simple "function checks" to ensure that the ac power supply is operational, the fuses or circuit breakers are not open circuit, the output current and voltage values are near to those anticipated and that the level of protection indicated by the permanent reference electrodes is satisfactory. These function checks are undertaken by the CP system's owner's staff on a weekly or monthly basis. Most systems require regular intervention to maintain them in a functioning status.

In addition, impressed current CP systems are intrinsically less reliable and more demanding in maintenance than galvanic anode CP systems. This is due to the greater complexity of the impressed current CP systems and as a result:

- less redundancy in anodes due to less anodes of higher current output being installed
- more risk of anode/cable connection and anode cable failures due to forced electrolytic failure of copper if there is any damage to insulation
- impressed current anodes often being less mechanically robust than cast galvanic anodes
- reliability issues with electrical and electronic power supplies and control systems.

The commissioning testing of an impressed current CP system will be more extensive than that for a galvanic anode CP system, in part due to the increased complexity of the components and also due to the need to undertake stray current (or interaction) testing. The latter is to ensure that the impressed current CP system (with higher individual current output anodes) is not causing high rates of corrosion to adjacent structures or vessels using the facilities being protected (see Section 2.2). However, the core of the performance verification will be as for galvanic anode CP systems:

- to demonstrate satisfactory distribution of cathodic polarisation and the electrical continuity between piles/elements of the structure (by steel/seawater potential measurements at different depths on many or all elements within a structure)
- to include "Instant OFF" potential measurements on coupons or elements of the structure to ascertain the extent of IR errors due to cathodic protection current flow in the measurement circuit (these can be significant with impressed current CP systems even in seawater).

During the maintenance/defects liability period it is common for the contractor to undertake three monthly repeat performance verification tests and adjustments of the operating parameters of the system in order to optimise and fully document the performance of the system.

Every year a complete survey of the impressed current CP system should be performed by a suitably qualified engineer (preferably in line with the operating and maintenance manual from the systems designers). The objectives are to evaluate the operation of the CP system and to make recommendations on any modifications or maintenance as required. After an inspection has been made, any maintenance (eg increase in output, for impressed current only, replacement of anodes, reference electrodes, etc) should be completed as soon as possible. All repairs to the system should be made in accordance with the original design or to any revision required to remedy performance or reliability defects.

If CP has been applied to an existing ALWC affected structure then the surface deposits in the MIC affected areas (in the intertidal low water zone) may become detached by the action of the protective current. This may reveal the extent of the clean, shiny/dull steel surface in this zone resulting from the previous MIC activity, although white calcareous deposits will normally be seen on the surface formed due to the action of the protective current. If the CP is operating correctly, then there should be little or no evidence of brown rust staining on such exposed bare steel surfaces below mid tide level.

3.1 Electrochemical monitoring of CP systems

Monitoring of CP systems is achieved by measurements of the electrochemical potential of the steel immersed in seawater with respect to a reference electrode placed close to the steel. The monitoring of CP systems should, generally, be in line with BS EN 13509:2003 "Cathodic Protection Measurement Techniques" and other relevant standards and recommended practices (see Section 2.2, Appendix 10).

The criteria for protection are detailed in BS EN 13174:2001 "Cathodic Protection for Harbour Installations" and BS EN 12473:2000 "General Principles of Cathodic Protection in Seawater". In general, for steel immersed in solutions which contain active SRB a steel/sea water potential more negative than -0.900 Volt (with respect to Ag/AgCl/seawater reference electrode) is determined as the protection criterion at which corrosion due to MIC will be controlled. If the structure is coated, a negative limit of -1.100 Volt (versus Ag/AgCl/seawater) is recommended to prevent overprotection and the resulting possibility of coating disbondment and/or hydrogen embrittlement problems of high yield steels (greater than $700N/mm^2$) or sensitive weld areas. BS 4515-1:2004 specifies maximum hardness values for welds of 275 HV10 for "sour service" or 300 HV10 or more for non-sour service. While SRB are unlikely to produce hydrogen sulfide levels which could be defined as "sour service", it would be prudent to work to the same maximum Vickers hardness value (ie 275 HV10).

Monitoring should be carried out at various locations of the structure in order to give a representative assessment of the CP system performance. The purpose of the survey is to measure potentials at locations where areas of the structure are likely to be less protected (less negative) than the protection criterion of -0.900 Volt (with respect to Ag/AgCl/seawater reference electrode). These areas include locations at a maximum distance from the anodes (seabed level) or locations of highest risk for ALWC (ie areas at or around LAT). The location of the reference electrodes is important as it dictates the steel/seawater interface at which the measurement is taken. The location of the steel contact for the measurement circuit is also important, the contact must be made at a location where there is good metallic, low resistance, connection to the steel/seawater interface at which it is intended to take the measurement.

Care must be taken not to place the reference electrode close to an anode during the surveys as the potential reading will be more negative (protected) than at locations remote from anodes.

All measurements are susceptible to influence by IR drop errors caused by CP currents flowing through the resistance of the seawater and the films on the steel surface generating a voltage (ie by Ohms Law) which is erroneously included in the CP potential measurement. The errors are usually small with galvanic anode CP systems but are more significant with impressed current CP systems due to greater individual anode current outputs. The use of "instant OFF" measurements by interrupting the dc output of impressed current CP systems or the use of steel coupons that are connected to the protected steel but that can be switched to open circuit for measurement purposes (for either galvanic anode or impressed current CP systems) may be beneficial for interpretation of a galvanic anode CP system performance and are required for impressed current CP system performance evaluation.

Permanently installed reference electrodes should be introduced and monitored at both the average and the extremities of area to be protected. Permanent reference electrodes are essential for impressed current CP systems operating in potentiostatic mode; since signals from them are used in the control system. If there is limited tidal variation and the impressed current system can be satisfactorily operated in galvanostatic (ie constant current) mode or for galvanic anode CP systems that are essentially self controlling, permanent reference electrodes may not be installed and the system performance can be monitored by portable reference electrodes.

Reference electrodes should either be Silver/Silver Chloride/seawater (Ag/AgCl/seawater) or a Ag/AgCl/0.5M KCl electrode, particularly in estuarine conditions where the potential of the former will be in error due to variations in chloride concentration in the water (in 100 Ohm cm water the apparent potential measured with a Ag/AgCl/seawater electrode will be some 30 mV in error; in such water the protection criterion value of -0.900 Volts will, therefore, be -0.930 Volts if measured with a Ag/AgCl/seawater electrode.

Zinc/seawater (Zn/seawater) electrodes are widely used for permanent electrodes but they are significantly inaccurate than the two reference electrodes cited above and can become more erroneous if used for long periods without cleaning.

All electrodes should be calibrated at regular intervals against a calibrated saturated calomel reference electrode (SCE) or another calibrated reference electrode. It is good practice to undertake such calibration before and after each survey and to declare the calibration data in the report of the survey.

Permanent reference electrodes should be calibrated by placing a portable reference electrode in close proximity. It is possible to install permanent data logging monitoring and remote control devices (SCADA or process controller based) on CP systems (particularly impressed current CP systems) which are able to continuously monitor outputs and protection levels. This can be used to limit the manpower commitment from structure owners/operators.

In order to have comparative results between the surveys throughout the service life of the structure, a potential measurement grid should be agreed between the owner, contractor and consultant. This ensures that the reference electrode placements between subsequent surveys are identical and allows comparison of surveys.

3.1.1 For galvanic CP systems

Typical galvanic CP systems specifically designed to prevent or remedy ALWC on steel structures are characterised by galvanic anodes placed at a level 1-2 m below MLWS which corresponds to the critical area for ALWC. It is necessary to inspect the system to ensure that there is both adequate protection and that the anodes will continue to give sufficient protection until their anticipated replacement. For deep water berths, it may be necessary for anodes also to be placed at a lower elevation. Sometimes anodes have been placed at alternate elevations in adjacent sheet steel pile in-pans, in order to achieve optimum distribution of current and levels of protection.

- in waters with adequate visibility, a careful visual inspection of the galvanic anodes with photography, if necessary, should be made every three to five years for:
    1. Evidence of mechanical damage to and loss of anodes and anode support rods or attachments.
    2. Average rate of anode metal consumption estimated by dimensional checks on the anodes and comparison with the manufactured dimensions. Anodes should be replaced when wastage exceeds their design utilisation factor which will be

about 80 to 90 per cent. The reason for any anode experiencing unusual consumption should be investigated and resolved.

- if permanent reference electrodes are installed, structure to electrolyte potentials should be read with high impedance multi-meters and recorded every one to three years
- measurements of the potential between the structure and a suitable reference electrode should be made initially every year to confirm satisfactory protection is being provided to all parts of the structure (see below). Protection is judged to be effective if the measured potential is more negative than -0.900 volt (versus Ag/AgCl/seawater). If the potential is less negative than this value, a detailed inspection of anodes should be performed. If the data are stable from one year to the next and if the anodes are securely welded to the structure, measurements can reasonably be extended to every one to three years
- for new CP installations, it is desirable to monitor the structure potential four times in the first year (and then once a year subsequently) as it can often take several months for the structure to achieve the specified protection potential even if satisfactorily designed
- if test coupons are installed they should be removed at the time of a detailed inspection and weighed to find the corrosion rate and hence confirm the effectiveness of the CP system
- environmental surveys for levels of Zn in the seawater and mud from bed level should be carried out. In particular on uncoated structures where the number of galvanic anodes is large and where the initial levels of Zn are already close to the limit given in the current UK annual average Environmental Quality Standard (EQS) for dissolved zinc in seawater (Matthiessen, 1999 and Bird, 1995).

3.1.2 For impressed current CP systems:

- a regular inspection regime is imperative to ensure that all components are functioning and that protective potentials are achieved at all times
- it is recommended that the system is switched off during visual inspection if divers are used to avoid any electric shock hazard. This applies to any diving in the vicinity of an impressed current system
- the following measurements should be recorded:
  1. Impressed current power source outputs on a weekly to monthly basis.
  2. Structure-to-electrolyte potentials at permanently installed reference electrodes and with a portable reference electrode on a three month basis.
  3. Full survey to include potential measurements, power source inspection (see point below) and stray current recordings on an annual basis.

- the inspection and examination of the impressed current CP system should also incorporate the entire CP system including the permanent reference electrodes, their cables, test connections to the steel and their cables, anodes and their cables, negative (drain point) connections to the steel and their cables and the DC power supply equipment and any data logging/remote control facilities, which should address the following:
  1. Operating conditions of DC power supply equipment.
  2. Mechanical and electrical integrity of all equipment.
  3. Functionality of the entire CP system.
  4. Maintenance of DC power supply equipment.
  5. Output current of impressed anodes.

6. Output voltage to achieve output current and calculated circuit resistance.
7. Insulation resistance of wiring.

- visual evidence for hydrogen embrittlement (ie cracking) of welds, fixings, etc and disbondment and/or blistering of any applied coatings indicative of overprotection should be completed during the diving survey every three to five years.

3.2 Electrochemical monitoring procedures:

- *structure to electrolyte potential monitoring;* measurements can be taken:
  1. From the surface by:
     - a portable reference electrode lowered into place and high impedance voltmeter
     - a permanently installed reference electrode and high impedance voltmeter

     If measurements are taken from the surface it is necessary to make an electrical connection from one terminal of the voltmeter to the steel structure.
  2. By divers with hand held potential monitoring devices.

- ***DC current measurements of impressed current CP system***; the DC current of individual anodes should be monitored either with a DC clamp meter or across shunts installed in anode junction/distribution boxes
- ***impressed current power source monitoring***; the DC voltage and current should be monitored at the outputs of the power sources. This can either be from permanently installed voltmeters and ammeters or by direct measurement with high impedance multimeter and clamp meter. Polarity of the dc output should be checked regularly since reverse polarity will result in enforced and rapid corrosion of the structure. Polarity reversal can occur by incompetent installation, commissioning or maintenance procedures
- ***stray current monitoring***; Adjoining or adjacent structures (including vessels berthed adjacent to a protected structure) to those structures with CP installed should be monitored to ensure that there is not a stray current problem. This is not often necessary for galvanic anode CP systems due to a large number of small current output anodes, however it is important for impressed current CP systems. Stray currents are detected by:
  1. Recording the structure-to-electrolyte potentials on the adjacent structures and protected structure prior to commissioning or with the CP system switched off.
  2. Commission or reconnect the CP system.
  3. Repeat the first step. A CP engineer should address any difference in the two potentials recorded.

Summary

- once remedial measures are complete, the structure should be regularly inspected to monitor their effectiveness against ALWC and any other deteriorating mechanisms that may be in operation
- the following inspection methods should be considered after one year and subsequently every three to five years:
    - *general* – visual inspection possibly with divers; consider video, photography, thickness measurements, on-site sampling etc (see Chapter 5 and Appendix 7)
    - *repair plates and welds* – visual and ultrasonic thickness measurements
    - *coatings* – visual inspection and repair, as necessary
    - *cathodic potection* – visual and electrochemical surveys. Electrochemical surveys should be carried out at least annually for impressed current systems and every one to three years for galvanic anode CP systems
- in order to generate comparative results between surveys, the owner, contractor and consultant should first agree a suitable inspection regime which should be adopted for all surveys. For example, in the case of potential measurements a grid matrix should be used so that the reference electrode is placed at identical locations for each survey
- the records of inspections, examinations and repairs should be accurately completed and preserved for reference in future inspections/surveys as part of the asset management cycle (see Chapter 4).

8 ALWC project procurement and management

Earlier chapters have presented technical solutions to the design, repair and monitoring of structures affected by ALWC.

This chapter covers the aspects of the procurement of the repairs and/or protection of existing structures that are specific in the context of ALWC and maritime works. The process for the procurement of new-build structures and their associated protective measures is standard and is, therefore, not included here.

8.1 Strategy for survey and works

Following the conclusions to initial inspections confirming that work will be required to repair and/or protect the structure, there are two typical strategies:

- detailed survey, followed by a separate works contract
- a single operation amalgamating a detailed survey and repair work.

Each approach presents advantages and disadvantages, as follows.

8.1.1 Separate survey and works contracts

Advantages

- the owner is more aware of the likely extent (and hence cost) of the repair and protection works prior to inviting tenders for the works contract and can, therefore, prepare a more detailed tender document and allow a smaller contingency than for the single operation strategy
- the survey may be completed by parties that are independent of the contractor who ultimately carries out the repair work
- more urgent repairs can be identified at an early stage and if funds are limited these works can be prioritised.

Disadvantages

- more than one mobilisation is required, so depending on the access method employed, it is likely that the overall cost will be higher
- it can be difficult for the survey to clearly designate areas of deterioration so that they will be easily identified during the later works contract
- it is possible, once started, that the works contract will identify additional areas of deterioration.

8.1.2 Combined survey and works contracts

Advantages

- only one mobilisation cost is incurred and the use of more extensive plant, such as a limpet dam, can be easily justified, thus allowing the survey and repair works to be undertaken in greater detail and with greater certainty.

Disadvantages

- the contract is let on the basis of the initial inspection. Hence the full extent of the work required is less certain and will not be fully apparent until the more detailed inspection undertaken during the contract. It is essential that experienced and competent operatives undertake the survey
- the survey undertaken during the works contract may be phased, for example undertaken as a limpet dam is moved along the structure. Hence the extent and the costs of work required will be revealed in an incremental fashion
- given the uncertainty of the extent of the repair work, a greater contingency should be allowed for in the budget costing. The extent of contingency allowance will depend on the detail of the initial (pre-contract) inspection.

8.2 Contractural and commercial issues

The following issues in particular require to be considered at the outset when compiling contract documentation for an ALWC project:

- the method of payment and form of contract
- supervision and quality control
- the effect of the works on port operations and vice versa
- prioritisation and contingency.

8.2.1 Method of payment and form of contract

The typical options for method of payment and type of contract for works of a significant financial cost, together with relevant contract forms, are illustrated below:

Method of payment	Type of contract	Typical forms
Price	Lump sum	• Engineering and Construction Contract 2nd edition, option A. • Institution of Civil Engineers 5th, 6th or 7th edition modified to suit. • Institution of Civil Engineers Conditions of Contract For Minor Works 3rd edition.
Price	Admeasurement	• Engineering and Construction Contract 2nd edition, option B. • Institution of Civil Engineers 5th, 6th or 7th edition. • Institution of Civil Engineers Conditions of Contract For Minor Works 3rd edition.
Cost	Cost reimbursable plus fee	• Engineering and Construction Contract 2nd edition, option E. • Institution of Civil Engineers Conditions of Contract For Minor Works 3rd edition.
Cost	Target cost	• Engineering and Construction Contract 2nd Edition, option C (with Activity Schedule). • Engineering and Construction Contract 2nd edition, option D (with Bill of Quantities). • Institution of Civil Engineers Conditions of Contract For Minor Works 3rd edition.

For works of a relatively minor financial scale the contract may settle simply on a written offer or quotation alongside a written acceptance or purchase order and where other peripheral conditions such as operation, safety and environment are catered for by separate permits to work or similar. In some instances, if the contractor is a one-stop-shop materials supplier and installer then it may be that firm's standard conditions of sale that could form the contract. The Institution of Civil Engineers Design and Construct may also be useful in this scenario if the works are likely to be of a larger financial scale. It contains useful provisions and conditions (such as insurance) which may make it worth considering as a basis regardless of project size. Both contractors and clients are mostly familiar with these items and time and disputes can be avoided by their use.

8.2.2 Selection of method of payment and contract form

The preferred method of payment and contract form would depend on the type of work to be undertaken and the certainty of the extent of the work. Payment by price is often favoured by the client because of the associated lower financial risk. However, if the work has to be undertaken urgently without the benefit of a procurement lead time, then payment by cost may have to be adopted.

It is a characteristic of ALWC that the extent of the associated repair works is not possible to define prior to letting a contract and hence an economical solution is unlikely to be achieved by a lump sum contract. However, installation of protection measures, such as application of coatings or installation of a CP system, is easily defined and can often be procured on a lump sum basis. Some contracts may call for the full scope of repair and protection and, in such cases, a balanced judgement on payment method will be required. Value engineering may be an option.

The true condition of steel piles below the water line can be notoriously difficult to predict, particularly with respect to holing. Unexpected holes can result in disproportionate additional expense on the works and it is critical to ensure the selected contractor has a sound strategy to deal with physical holes and that there is an adequate mechanism in place for accurate recompense.

Notwithstanding the numerous methods of forming a contract, the best form of contract is that which is appropriate to the scale and complexity of work to be done, and which is acceptable to both parties.

8.2.3 Supervision and quality control

Undertaking works in the maritime environment is sometimes awkward in terms of access, tides, shipping and weather and it is essential that the required standards of workmanship are achieved at the first attempt. Quality control is best facilitated when inspection and assessment are undertaken by an experienced, competent engineer and it is up to the client whether this is done independently of the contractor, internally or by a representative. While this will involve additional expense to the owner of the structure it will provide additional confidence that the required repairs have been undertaken satisfactorily.

Under some circumstances, such as design and installation of a CP system, a suitable guarantee of performance from a reputable supplier or contractor may be a reassuring alternative but again a premium may be involved pushing the project cost up further.

The method of working will determine the practicality of how inspection work can be undertaken. Work in the dry using a limpet cofferdam or similar would allow land-based staff direct access to the work face, whereas work in the wet using divers would, if the owner's staff are not suitably qualified, require the use of another, independent, diving organisation or the use of video and telecommunication links.

Given the disparity of working methods available, for example work in the wet using divers or work in the dry using a cofferdam, the standards of workmanship and the level of inspection that will be imposed should be decided prior to issuing, and clearly stated in, the tender documents. It may be prudent to have test fabrications or coatings specified for the outset of the works, eg to be performed during high tide at mid tide level, for inspection and agreement of standards by all parties concerned before work starts in earnest.

8.2.4 Prioritisation, contingency and scope of work

Given the nature of the work, it is expected that further areas of deterioration become clear after the contract has been let. It is, therefore, advisable to:

- allow a contingency, so that there is scope for additional work to be undertaken during the contract and not delayed so that the additional mobilisation costs associated with a second contract are incurred. This can be achieved by agreeing a day rate or hourly rate for extra works or, alternatively, using a schedule of rates for the agreed elements of the works
- wherever possible determine in advance those works which are critical, so that in the event that additional work is identified or issues such as weather restrictions limit the extent of the work that can be progressed, there is a clear understanding of which operations should be given priority
- allow for extra thickness readings and, if not already part of the scope of work, steel patch plating should defects become apparent in the course of the works. Removal of existing features such as fenders or old anodes should be included, if applicable. It might be useful to take advantage of the mobilisation to carry out repairs on items such as ladders or fenders.

8.2.5 Interaction with normal facility operations

The owner has to determine whether the repair/protection work will conflict with normal facility operations and, if so, which will take priority. Disruption of the contractor's operations will have cost implications but this may be less than the cost of disruption to operational services. If there is potential for disruption to the works from facility operations, the following need to be determined and incorporated in the contract documentation:

- line of communication between the facility operator and the contractor
- the expected level of disruption
- contractor's operations that would and would not be permitted to continue during facility operations, noting that in the case of ports, for example, many operations cannot be halted at short notice
- the notice of disruption the owner will give to the contractor
- the method of any payment owed to the contractor in respect of disruption of operations (the owner must decide whether tender prices are to include risk of delays due to facility operations or other causes such as bad weather).

8.2.6 Programming and urgency

The contract documents should consider the timescale within which the work is required to be undertaken, having due regard for the interaction with normal facility operations and other issues that may delay the contractor's operations, including weather. Given the nature of the work the document should be realistic in terms of time-scale, because a requirement for increased speed will almost always result in increased cost. Nevertheless, for operational reasons, it may be cost/time effective to set up a 24 hour operation to maximise work and minimise demurrage. This optimises use of expensive plant and may be competitively priced if it suits the contractor to take on other work by finishing sooner. Alternatively, for the same reason, it may be acceptable to consider multiple teams and plant.

8.2.7 Invitation of tenders

There are three methods to select the contractor:

- open tender
- invited tender
- negotiated contract.

Given the specialised nature of ALWC repairs and the distinctive environment in which they need to be carried out, it is suggested that only contractors with proven experience or ability should be invited to tender. There are merits, therefore, to holding an open prequalification process, from which a select list of suitable tenderers can be chosen, if not an open tender process. Assistance with the selection of suitably qualified contractors may be sought from independent consulting engineers.

Legislation regarding procurement processes for works above a given threshold value must be complied with, although many repair works contracts fall below current thresholds.

Adoption of a negotiated tender removes the element of competition. In exceptional circumstances, such as when repairs should be put in place with urgency, there may be benefits to negotiating a contract with a selected firm of known integrity and ability.

Note that projects that are grant assisted may have specified tendering procedures.

8.2.8 Tender and contract documentation

Irrespective of the form of contract and method of payment adopted, the tender and contract documentation should include:

- instructions to tenderers. Included should be an express requirement for tenderers to visit the site to fully appreciate the intricacies that may be involved
- form of tender
- the conditions of contract
- conditions of particular application – additional to the main contractual conditions to ensure the contractor complies with the owner's specific requirements and constraints, including requirements in respect of planning, port operations, health and safety, environmental issues, local/port bye-laws, applicable statutory and common laws and working language for international contracts, etc

EAGLE LIBRARY
The University of [illegible]ton
Eagle Campus

- a detailed technical specification to describe the standards of materials and workmanship or the level of materials performance required
- quality documentation, as may be required, should be clearly stated.
- other information provided by the contractor should be stated, such as details (CVs) of the operatives proposed for the job as demonstration of their competence and experience
- Bill of Quantities, or equivalent depending on the form of contract – to delineate the extent of the work (and allowance for additional work) and to ensure like for like comparison of tenders. Ensure a section for insertion of method related charges by the tenderer who, by their expertise, may be aware of items that need to be allowed for not otherwise covered in the bill
- drawings – to detail the extent of the work, including the remedial measures which are to be installed (eg size and extent of welds around strengthening plates). On existing structures drawings should be developed to portray the true condition of the quay and piles as this can have a significant effect on the design of both temporary and permanent works
- pre-tender Health and Safety plan in accordance with CDM regulations and other safety information and requirements
- form of agreement – the legal undertaking entered into by the owner and the contractor.

Tenderers should be required to provide details of their proposed methods of working, together with an indicative programme, to ensure they have correctly understood the requirements and will not conflict with the owner's other operations. The tenderer is asked to provide an outline resourcing schedule that defines their commitment to the project in alignment with the indicative programme. The owner's requirements should also clearly define the contractor's responsibility for any design work which they are required to undertake, for example as would be required for a CP or proprietary repair system. Box 8.1 shows a potential "checklist" for providing information to tenderers in addition to the above.

Box 8.1 Provision of information to tenderers

In order for the tenderer to accurately and fairly price the required works, owners should provide all information available from written, recollected, photographic and drawn records, for example:

- *as-built drawings, construction photographs, pile sections, pile material and grade as specified and/or as supplied*
- *all modifications since construction – details of any previous repairs and protection systems (eg CP, coatings, etc)*
- *paint specification with manufacturer's technical data and health and safety sheets for all materials supplied*
- *fenders and ladders*
- *results of any surveys and previous inspections including low water and diving surveys, thickness readings on steel and paint coatings and photographs*
- *current condition of structure from:*
  - *ideally a photographic survey carried out at very low spring tide, of the whole structure showing capping to low water level and detailed photographs identifying any holes*
  - *general arrangement drawings of quay surface including any drainage, stats and services, bollards or manholes*
  - *dredge level*

- ◻ *cross-section showing toe depths and tidal range*
- ◻ *information on twisted piles, split clutches, ship damage, previous plating and records of any voiding or quay surface collapses*
- ◻ *operational constraints such as undivertable shipping, quay possession, and environmental restrictions.*

The technical specification, especially for design and construct contracts, should contain as much detail as possible on:

- *expected design life (and how "life" is defined), and life to first maintenance*
- *relative importance of whole life cost and capital cost*
- *level of inspection and interim maintenance expected on completion of the works*
- *how and by whom the extent of work will be identified- who will carry out a survey and at what point?*
- *is all work to be completed out in the dry or is diver welding acceptable? To what specification?*
- *detailed specification and methods and testing specification for remediation and protection systems:*
    - ◻ *CP; eg salinity readings at different states of tide, design code to be used, protection potential to be achieved, whether impressed current CP is acceptable (maintenance issue) or will galvanic CP be stipulated?*
    - ◻ *plating eg type (flat or bent), grade, height and thickness, whether anodic to parent, minimum parent metal thickness before plating, arrangements for odd size and position plates*
    - ◻ *coating eg extent, methods of surface preparation and application, thickness and type, cathodic disbondment*

8.2.9 Selection of contractor

It is advisable that contractors and any proprietary methods of working are pre-qualified before tender. The selection process should take account not only of price, but also of the tenderers' understanding of the contract requirements, including the job site, methods of working, resourcing, programming, understanding of and compliance with Health, Safety and Environmental issues.

Highlight in the tender documents how tenders will be evaluated will both benefit the tenderer and enable the owner to evaluate their priorities and requirements. A scoring system should include weighting applied to each category, such as:

- tenderer's track record, experience and financial standing
- health and safety and environmental considerations
- cost, including an estimate of the level of supervision required
- system proposed, including compatibility with operational constraint and how the quality achieved can be demonstrated (particularly important where divers are used)
- programme.

It is particularly important that tenders understand the extent of interaction with port operations and the subsequent procedures. The owner, or representative, should meet with the favoured tenderer(s) prior to award of the contract to ensure that all requirements have been fully covered in the tender price, and that qualifications to the tender, if any, are clear and acceptable.

9 Areas requiring further research

The reader of this report may already be aware of, or have noticed several anomalies surrounding the ALWC phenomenon. Many questions still remain unanswered including:

- why has there been such an apparent rapid spread and growth of ALWC?
- will these trends continue or even accelerate?
- what are the actual controlling factors that determine susceptibility to ALWC?
- how far can ALWC extend into the continuously immersed zone and what factors controls this?
- does ALWC attack all types of steel?
- can displacement or removal of surfaces deposits in the susceptible zone initiate or exacerbate ALWC attack?
- why does ALWC have a general tendency to attack at certain locations for particular pile geometries?
- why do structures located in different areas within an affected installation appear to show such a wide variation in susceptibility to ALWC?
- why do installations in the same geographic area appear to exhibit such a widely varying susceptibility to ALWC?
- are conventional protection systems, perhaps with a track record in other applications, able to deal effectively with the ALWC problem?
- will new treatment systems be developed specifically to treat ALWC in a more focused way and will they work? If so, what are their limitations?
- how can designs be improved to reduce or avoid the likelihood of ALWC on marine structures?

9.1 Research topics

There are a number of avenues where future research in the understanding, identification and protection against ALWC might be usefully directed and these are outlined below. Some have previously been investigated but the results have been inconclusive. For these and the remaining areas, further research will be necessary, likely requiring substantial investments in order to obtain satisfactory outcomes. The research will need to be specifically targeted to ALWC as a special case of the more general MIC condition, the mechanisms of which are still not fully understood. Other more general areas requiring fundamental research include whether crevice corrosion is likely to be a problem due to inadequately welded repair plates and the need for sealing resultant voids between the edge of the repair plate and the substrate steel.

9.1.1 Mechanism(s) for ALWC

An initial understanding on mechanism has been established from previous research studies (see Appendix 3) but there is still a fundamental lack of knowledge on the precise mechanism(s) of ALWC (see Chapter 2, Section 2.3.3). A focused, more fundamental research programme is required to establish the mechanisms of:

- the initiation process of ALWC and the factors that affect the incubation time prior to its full development. This will identify why ALWC gets established and may permit local environmental alteration to prevent it

- the propagation process of ALWC once established and the factors that affect the corrosion rate. This will identify a likely asset lifetime and, importantly, allow the application of effective, in-situ, remedial measures that will prevent re-occurrence.

In the absence of a known mechanism for initiation and propagation of ALWC that is reproducible under laboratory conditions, any evaluation can only be undertaken using large-scale and costly field trials over an extended number of years. A reliable laboratory simulation environment where the local conditions at an ALWC patch may be reproduced effectively would allow reproducible studies on protection strategies to be developed in a cost-effective manner prior to large-scale field trails. The establishment of a mechanism for ALWC would also help in deciding on the most effective protection strategy and how it should be applied (eg which zones require to be coated). It would also assist in the development of other solutions to the problem.

9.1.2 Protection of steel against ALWC

1. Conventional systems

There is a requirement to investigate the practical methods of protection against ALWC including:

- corrosion protection applied prior to installation: eg cathodic protection design criteria (ie protection potentials, etc); pre-applied coatings sufficiently thick and mechanically sound to resist installation stresses including the driving of piles
- corrosion protection added after installation: eg design current density criteria for reliable installation of CP systems on existing structures (see Appendix 10, Section 2.2 b); in-situ application/ installation of coatings, linings and wrappings.

2. New treatments

Most of the published work to date concerned with the treatment of and protection against ALWC describes the use of conventional systems, some with a track record in related or in other applications. There is, therefore, a requirement to monitor such systems as part of a research programme to establish or confirm their overall effectiveness.

Since ALWC would appear not to be inevitable and when it does occur it only typically affects a small percentage of a structure's surface, there is also a requirement to develop new treatments which can reliably treat the problem as it occurs, in a more focused way (see Section 9.2). To be acceptable for use at a busy maritime installation, any such treatments will be required to:

- produce minimum disruption to installation activities while the treatment is being implemented
- be environmentally friendly
- need only a relatively short time to implement
- be cost effective, ideally significantly cheaper than existing protection systems
- be low maintenance or maintenance free.

9.1.3 Identification of ALWC

There is a need to develop (or modify existing) techniques that can quickly and easily identify which individual elements of a structure are suffering from ALWC. For example, use of:

- NDT methods to determine the condition of individual structural elements without the need to remove surface deposits (which may be protective – see Section 5.2.5) and/or only that require access to the upper exposed steel surfaces to be applied (reducing the need for close-up surveys and/or divers). Candidate methods for evaluation include pulsed eddy current systems and guided wave ultrasonics
- electrochemical techniques such as cathodic cleaning to preferentially remove the less well adhered surface deposits at ALWC patches using hydrogen generated on the surface as a result of the applied current. This technique is reported to be effective, under controlled conditions, as part of a new process under development for the treatment of ALWC (Marshall, 2002).

9.1.4 Monitoring of ALWC

The possible use of instrumented, structural elements incorporating monitoring devices (eg flush mounted corrosion monitoring probes, low cost ultrasonic thickness transducers, sessile bacterial monitoring probes, etc) which could be installed either at new build sites or existing ALWC affected sites. As with all monitoring devices their effectiveness is dependant on their location relative to where the problem is occurring. Due to the known random nature of distribution of ALWC within installations this may be a particular difficulty.

9.1.5 Assessment of ALWC affected structures

Methods need to be developed for more accurate forecast of structural redundancy in design life calculations from consideration of specific geometry of steel loss locations. It is evident that most steel structures do not fall down when the corrosion allowance has been exceeded, or even if the structures become holed.

9.1.6 Predictive model

A statistical model has been proposed aimed at quantifying the risk of a structure being affected by ALWC in a given environment and operation (see Appendix 3). However, further work is required to develop a more reliable predictive model that would greatly assist in making an informed decision on the most appropriate protection strategy for any new build structure.

9.2 New developments

A new process which is currently under development for the treatment of ALWC in a specific rather than general way has been reported (Marshall, 2002). The patented treatment method (a three stage electrochemical process) and apparatus (modular mesh electrode system) allow for in-situ cleaning, sterilisation and coating at ALWC patches, all active ingredients being derived from the seawater itself (Breakell, 1999). An optional initial identification stage is also being researched to locate ALWC patches that have not yet sufficiently developed to allow visual characterisation.

10 Recommendations for good practice

This chapter, along with the executive summary and the summaries at the end of Chapters 1-7, presents guidance on general good practice in dealing with ALWC. Following the guidance given in this report and complying with any legislation will help owners and their technical and commercial management to demonstrate that they have intended to meet their legal responsibilities for safe operation of their port or other maritime infrastructure.

Owners' responsibilities

- owners should recognise the rapid deterioration effect of ALWC and its potential detrimental impact on both new build and existing structures
- owners should consider their technical, commercial and legal responsibilities for managing operational, environmental and Health and Safety risks. Failure to implement appropriate measures for the monitoring and control of corrosion can expose an organisation and its management to the full severity of Health and Safety and the Environmental Legislation if a fatal incident, pollution release or harm to the public occurs
- encourage design, inspection or maintenance solutions to adequately take into account the possibility of ALWC occurring and the need for adequate and safe access. The CDM Regulations impose a duty on clients, designers and contractors to have adequate regard to the possible effects of design on the hazards and risks which may arise to those constructing, cleaning, maintaining and demolishing the structure as well as others who might by affected by their activities. It is also a requirement that at the end of construction that a Health and Safety File is produced. This File should include the inspection and anticipated maintenance and any redundancy built into the design.

Asset management

- implement an asset management system to improve health and safety management, increase economical efficiency and optimise expenditures
- maintain ongoing condition appraisal documentation for each structure. It is important for ALWC detection to include exposed surfaces in the intertidal low water and continuously immersed zones. Analyse to predict when maintenance and repair will be needed as part of the asset management programme
- be aware that early intervention by way of regular condition appraisals, repair and protection strategies will make long-term economic sense
- when external support is required, always use specialised resources and services with previous experience of ALWC.

New build structures

- for new build structures include CP and coatings for corrosion protection in any design
- it will help facilitate future analysis if the original steel section thickness and, ideally, the minimum thickness of section required to carry design loads are clearly shown on design drawings
- operation and maintenance manuals for new build structures should recognise the risk of ALWC occurring and stress the importance of ongoing monitoring of condition in the susceptible zones and give the required inspection intervals.

Existing structures

- for structures affected by ALWC, seek independent, professional advice to give guidance on the most appropriate and cost effective corrosion protection strategy and when to apply it
- provide all relevant information concerning a structure to maintenance and repair contractors ensuring that this is fully taken into account
- regularly monitor and maintain any installed corrosion protection system for condition and performance throughout the service life of the structure and system.

Further research

- make funding available to allow necessary R&D work:
    1. To obtain more definitive information on the scale of the ALWC problem within the UK.
    2. To elucidate the mechanism of ALWC.
    3. To develop a reliable predictive model for ALWC occurrence.
    4. To encourage the development of new prevention, inspection and remedial treatment methods designed to treat the ALWC problem in a more specific rather than general way.

11 REFERENCES

AMERICAN IRON AND STEEL INSTITUTE (1981)
"Handbook of corrosion protection for steel pile structures in marine environments"
Ref. PI 072-581-5M-EP

ANON (2003)
"Safe and sound defence from flexcrete"
Port Engineering Management Jan/Feb 2004, Vol 22, p 17 also "ALWC", November 2003

ARUP, H H and GLANTZ G (1963)
"An evaluation of corrosion of sheet steel piling in Danish harbours"
Proceedings of the 2nd International Congress on Metallic Corrosion

ASTM D610
"Standard test method for evaluating degree of rust on painted steel surfaces"
American Society for Testing Materials, 1968

BAILEY G (2004)
"Assessing risk of microbial corrosion on berths and jetties"
Marine and Ports Review, 2004, pp 47-48

BALDWIN, R H (1998)
"Black powder in the gas industry – sources, characteristics and treatment"
Report No. TA 97-4, Gas Machinery Research Council, May 1998

BALMER, B (1984)
"Underwater inspection for corrosion" *Corrosion and Marine Growth on Offshore Structures*, edited by J R Lewis and AD Mercer, 1984, ISBN 0-85312-564-3

BARREDA, H, HUBERT, C and FARDELLA, C (2002)
"Cathodic protection and associated coatings"
European Federation of Corrosion Event No 254 (June 2002); paper reference; "Systemes de Protection Anticorrosion Duplex de Structures Metalliques a la Mer"

BEECH, I, CAMPBELL, S, MILLS, G and WALSH, F (1996)
"Engineering problems caused by microbial corrosion and their prevention"
Corrosion Management, June/July 1996

BIRD, P, COMBER, S D W, GARDNER, M J and RAVENSCROFT (1996)
"Zinc inputs to coastal waters from sacrificial anodes"
The Science of the Total Environment 181, 1996, pp 257-264

BOOTH G H, ROBB J A and WAKERLEY D S
"The influence of ferrous ions on the anaerobic corrosion of mild steel by actively growing cultures of SRB"
3rd International Congress on Metallic Corrosion, Moscow, 967, Vol 12, Section 5, p 542

BREAKELL, J E (1999)
"ALWC corrosion treatment method and apparatus"
Patent Number GB2356405, 1999

BREAKELL, J E (1998)
"Literature review of LAT corrosion"
Mott MacDonald Ltd. Project No. 48859/01/A, Sept. 1998

BRITISH STANDARD 3900: Part H3: 1983
"Designation of degree of rusting"
British Standard Methods of test for paints

BULTMAN, J D and SOUTHWELL, C R (1977)
"Biocorrosion of structural steels in seawater"
Reviews on Coatings and Corrosion 1977, Vol 2, pp 187-214

CAMPBELL, S A, SCANNELL R A and WALSH F C (1990)
"Microbially assisted pitting corrosion of ships hull plate"
Industrial Corrosion, 1990, Vol 8, No1, pp 7-14

CARMICHEAL, J (1998)
"ALWC – an introduction"
Arch Henderson & Partners, Accelerated Low Water Corrosion, ICE London, October 1998

CHEUNG C W S, WALSH F C, CAMPELL S A, CHAO W T and BEECH I B (1994)
"Microbial contributions to the Marine Corrosion of Steel Piling"
International Biodeterioration & Biodegradation, 1994, pp 259-274

CHRISTIE, J (1999)
"Steel pile corrosion survey model – inspection procedures for accelerated corrosion on marine steel piles"
Port Technology International, 11th Edition, 2000

COPENHAGEN, W J (1934)
"Sulfur as a factor in the corrosion of iron and steel structures in the sea" J Trans. Royal Society of South Africa, 1934, Vol. 22, No. 2

COPENHAGEN, W J (1954)
"Sulphur as a factor in the corrosion of iron and steel structures in the sea"
South African Industrial Chemist 1954, Vol 8, pp 32-35

CUNNINGHAM, A P (1998)
"Coating systems for steel sheet piling"
Protective Coatings Europe February 1998

DOUGHTY, A S, JOHNSON, J B, RICHARDSON, J A, SCANTLEBURY, J D and WOOD, G C (1980)
"Protection of metals in ports and harbours: research and findings"
Building Research Establishment Report, 1980

ELLIS, S H (1914)
"Corrosion of steel wharves at Kowloon"
Proc. Inst. Civil Eng., 199, 133, Paper No. 4090, 1914

EMAMI, N K, NICHOLSON, A, WREN, J and MONCARZ, P D
"Micro-biological attack on concrete – a threat to concrete infrastructure"
International Conference on Forensic Engineering, A Professional Approach to Investigation, ICE London UK, September 28-29, 1998, Thomas Telford Publications

FUKUTE et al (1990)
"Steel structures in ports and harbour facilities – actual conditions of corrosion and counter measures"
Bulletin PIANC-AIPCN 1990, No. 68, pp 93-110

GEHRKE, T and SAND, W (2003)
"Interactions between micro-organisms and physicochemical factors cause MIC of steel pilings in Harbours (ALWC)"
Corrosion 2003 Paper No 03557

GENIN, J M R, OLOWE A A and RESIAK B (1993)
"MIC of steel sheet piles in a harbour and the presence of green rust two compound"
Progress in the understanding and Prevention of Corrosion, European Corrosion – European Federation of Corrosion Services, Vol. 11, Proc. Conference, Barcelona, July 1993

GRAFF, M (2002)
"Accelerated Low Water Corrosion (ALWC) and Microbial Induced Corrosion (MIC): two sides of the same coin?"
Accelerated Low Water Corrosion Conference, Denmark, 2002

GUBNER, R and BEECH, I (1999)
"Statistical assessment of the ALWC in the marine environment"
Corrosion 99, Paper No 318

GUPTA, N K
"Bio invasion through ballast water"
Port Technology International Edition 13, Section 5, June 2001, pp 75-76

HEALTH & SAFETY COMMISSION (2001)
"Managing health and safety in construction: construction (design and management) regulations 1994. Approved Code of Practice and Guidance"
HSG224, HSE Books, Sudbury

HODGSON, M (1998)
"Accelerated low water corrosion – a civil engineer's viewpoint"
Corrosion Management September/October 1998

HOLLEYWOOD, J and PRIMROSE, J D (1978)
"Systematic maintenance and inspection of a major port"
Maintenance of Marine Structures, ICE 1978

HOU, B R, ZHANG J, DUAN J Z, LI Y and ZHANG J L (2003)
"Corrosion of thermally sprayed zinc and aluminium coatings in simulated splash and tidal zone conditions"
Corrosion Engineering, Science and Technology 2003, Vol 38, No 2

HUMBLE H A (1949)
"The cathodic protection of steel sheet piling in seawater"
5th Annual Conference NACE Cincinnati Ohio April 11-14, 1949

INCO DATABOOKS (1973)
"A guide to selection of marine materials"

JOHNSON, K et al (1997)
"Low water corrosion on steel piles in marine waters"
ECSC Final Report EUR 17868, 1997 (ISSN 1018-5593)

KENNEDY, E R and WILSON, J S (1967)
"New York Harbour Corrosion – Port of New York Authority Conducts Pile Survey"
Materials Protection January 1967 Vol 6 pp 53-56

KING, R A and MILLER, J D A (1989)
"Cathodic protection and sulphate reducing bacteria"
UK Corrosion, Blackpool, 1989

KIRK, W W and PIKUL S J (1990)
"Seawater corrosivity around the world: Results from three years of testing"
Corrosion in Natural Waters, ASTM STP 1086, 1990, pp 2-36

KUMER, A and STEPHENSON, L D (2002)
"Accelerated Low Water Corrosion of steel pilings in seawater"
PIANC 2002, 30th International Navigation Congress, Sydney, 2002

LEE, W, LEWANDOWSKI, Z, NIELSEN, P H, HAMILTON, W A (1995)
"Role of Sulphate Reducing Bacteria in Corrosion of Mild Steel – a review"
Biofouling, 1995, Vol 8, pp 165-194

MARSH, E (1998)
"Protection methods of steel sheet piling – UK experience"
British Steel, Accelerated Low Water Corrosion, ICE London, October 1998

MARSH E (2003)
Corus, private communication

MARSHALL, D and BREAKELL, J E (2002)
"The development of an innovative treatment for dealing with Accelerated Low Water (ALWC) problems in ports and harbours"
ALWC Conference, Denmark, 2002

MATTHIESSEN, P, REED, J and JOHNSON, M (1999)
"Sources and potential effects of copper and zinc concentrations in the estuarine waters of Essex and Suffolk, United Kingdom"
Published by Elsevier Science Ltd., 1999

MCNEIL, M B and LITTLE, B J (1990)
"Mackinawite formation during microbial corrosion"
Corrosion, July 1990, Vol. 465, No. 7, pp 599-600

MELCHER, R E (1997)
"Modelling of marine corrosion of steel specimens"
Corrosion Testing in Natural waters: 2nd Volume, ASTM STP 1300, ASTM, 1997

MELCHERS, R E (2003)
"Modelling of marine immersion corrosion for mild and low-alloy steels – Part 1: Phenomenological model"
Corrosion Vol. 59, No 4, April 2003, pp 319-334

MELCHERS, R E (1994)
"Pitting corrosion in marine environments – a review"
Research Report No 101.07.1994 The University of Newcastle New South Wales, Australia, 1994

MORELY, J et al (1978)
"Corrosion of SSP at Southampton docks"
British Steel Corp. Report No. T/CS/906/4/78/C, Feb 1978

MOULIN, J M (2001)
"Prevention of accelerated low water corrosion on steel piling structures due to microbially influenced corrosion mechanisms"
ECSC Final Report EUR 20043, 2001 (ISBN 92-894-2068-5)

MOULIN, J M et al (1995)
"Special corrosion of SSP in temperate seawater"
NACE International Conference on MIC, May 8-10, 1995, New Orleans, Louisiana, USA

MOULIN, J M (1999)
"Protection methods of steel sheet piling against ALWC"
EUROCORR, 1999

NEW CIVIL ENGINEER (March 1998)
"Steel eating bugs spread worldwide"
March 1998

NEW CIVIL ENGINEER (October 1998)
"Biocorrosion threat to UK coastal defence"
October 1998

NEW CIVIL ENGINEER (2003)
"Beside the sea side"
18 Sept 2003

THE OVERSEAS COASTAL AREA DEVELOPMENT INSTITUTE OF JAPAN (1998)
"Corrosion protection and repair manual for port and harbour steel structures"
1998

PHULL, B S, PIKUL, S J and KAIN, R M (1997)
"Seawater corrosivity around the world: Results from five yeas of testing"
Corrosion Testing in Natural Waters: Second Volume ASTM STP 1300 American Society for Testing and Materials, 1997

PIANC (2004)
"Inspection, maintenance and repair of maritime structures exposed to damage and material degradation caused by a salt water environment"
PIANC Report, Working Group 17-2004

REINHARD, K (1998)
"German experience"
Hoesch Spundwand und Profil, Accelerated Low Water Corrosion, ICE London, October 1998

ROSS, C W (1948)
"Experimental steel pile groins, Florida"
US Army Corps of Engineers, Beach Erosion Board Technical Memo No. 10, 1948

SANDBERG, B (2002)
"Cathodic protection of steel sheet piling in brackish water"
Swedish Corrosion Institute, Sweden, Accelerated Low Water Corrosion Conference, Denmark 2002

SANDISON , A AND PERRY B
"Accelerated corrosion of sheet pile quay walls – a cost effective repair solution"
NACE 2001

SCHWINDT, J and BAUMANN, M (1999)
"Development and testing of tar-free polyurethane coatings for cathodic-protected sheet pilings"
Protective Coatings Europe, July 1999

SEMERAD, V A W, CORSIGLIA, F A, WEAVER, D and COX, G (2003)
"Testing of epoxy adhesives for splash zone coating retrofit system for marine pipeline riser applications"
Corrosion 2003, NACE Paper No. 03042

SHAW, M (1998)
"Diving inspections and corrosion below lowest astronomical tide" Kirk McClure Morton, Accelerated Low Water Corrosion – Heriot-Watt University, September 1999 Not in

SHELLEY, T (1998)
"ALWC experience in piled structures"
Port of Felixstowe, Accelerated Low Water Corrosion, ICE London, October 1998

SHELLEY, T (2002)
"ALWC survey of UK installations, 2001"
ALWC Conference, Denmark 2002

SMART, J S (1997)
"An introduction to microbially influenced corrosion in the marine industry"
Journal of Protective Coatings & Linings, February 1997, pp 56-64

SMITH, M, BOWLEY, C and WILLIAMS, L (2002)
"In-situ protection of splash zones – 30 years on"
Winn & Coales (Denso) Ltd., NACE 2002, Denver, USA

SONCK, W (2002)
"Accelerated Low Water Corrosion"
Acotec NV, Belgium, ALWC Conference, Denmark 2002

SOYANG, T (1998)
"Underwater corrosion of sheet piling"
Acotec N V, Accelerated Low Water Corrosion, ICE London, October 1998

STEELE, G (2002)
"Accelerated low water corrosion experiences from the port of Felixstowe and port of Southampton"
Associated British Ports, Accelerated Low Water Corrosion Conference, Denmark 2002

STOTT, J F D (1998)
"Corrosion assessment and control of MIC"
Corrosion, 1998, p 224

TILLER, A K (1983)
"Electrochemical aspects of microbial corrosion; and overview"
Microbial Corrosion, NPL Teddington, March 1983

TSUCHIDA, H and YOKOI T (1985)
"Studies of the corrosion of steel materials in a marine environment"
26th International Navigation Congress, Brussels, 16-22 June 1985

TSURTA, S and ZEN, K (1973)
"Engineering aspect in corrosion prevention of steel structures in offshore regions"
Proc. 23rd Congress of PIANC, 1973, pp 133-156

VAN DAMME, L (2002)
"Belgian system against ALWC"
Ministry of the Flemish Community, Belgium, Accelerated Low Water Corrosion Conference, Denmark 2002

VAN DAMME, L and VRELUST, W (1999)
"Low water corrosion of steel pilings"
Bulletin, International Navigation Association, Belgium, ISSU 101, 1999 pp 47-52

VESTH-HANSEN, J (2002)
"Routine inspection and protection of steel sheet piles in the port of Copenhagen"
Accelerated Low Water Corrosion Conference, Denmark 2002

VRELUST, W, VAN DRAEGE, E and VERBEECK, A (1992)
"The treatment of steel sheet piles with anti-corrosive means"
10th International Harbour Congress 1992, Vol 10

WINDSOR, B (2000)
"Underwater corrosion protection – resisting attack by the microbial metal munchers"
RUST 2000, US Navy Information Exchange Louisville, Kentucky, 5-8 June 2000

WYATT, B S (1999)
"Cathodic protection: A solution for Accelerated Low Water Corrosion"
Corrosion Control Services, Accelerated Low Water Corrosion, Heriot-Watt University, September 1999

ZEN, K (1985)
"Corrosion and corrosion control of steel structures under marine environment"
4th Asian-Pacific Corrosion Control Conference Material Conservation and Corrosion Control Volume 1 Tokyo Japan, 26-31 May 1985

APPENDICES

A1 CORROSION REACTIONS OF STEEL IN SEAWATER

The corrosion reaction of steel in seawater is an electrochemical process involving a flow of electricity between certain areas on the metal surface (known as anodes and cathodes) through the seawater (the electrolyte solution) which is capable of conducting electric current. The term anode is used to describe that portion of the metal surface which is corroded and from which metal ions leaves the surface to enter the solution generating electrons that pass through the metal to the cathode. The term cathode describes the metal surface from where electrons leave the surface to take part in a reduction reaction. This is commonly referred to as a corrosion cell.

For corrosion to occur the following should be present:

- an anodic reaction (sometimes referred to as "oxidation", loss of electrons)
- electron transfer (electric current)
- a cathodic reaction (sometimes referred to as "reduction", gain of electrons)
- ionic transfer (seawater contact).

If any one of the four items is either absent or can be controlled/reduced then the corrosion process itself can be halted/slowed down.

For corrosion to occur the anodic reaction and the cathodic reaction must take place simultaneously on the metal surface at the anodes and cathodes respectively. At the anode metal ions, positively charged atoms, leave the surface and enter into solution (cations). For example in the case of iron (the principal constituent of steel) the anodic reaction will be:

$$2Fe \rightarrow 2Fe^{2+} + 4e^- \quad (1)$$

The electrons generated in this reaction flow through the metal to complete the circuit to the cathode where a corresponding reaction (the cathodic reaction) consumes these electrons thereby restoring the electrical imbalance caused by reaction (1). The most common cathodic reaction on steel in unpolluted, aerated seawater is oxygen reduction (derived from dissolved oxygen in the seawater)

$$O_2 + 2H_2O + 4e^- \rightarrow 4OH- \quad (2)$$

Corrosion products may appear in the form of deposits or scales by precipitation when anodic and cathodic products react with each other:

$$2Fe^{2+} \text{ (from the anode)} + 4OH^- \text{ (from the cathode)} \rightarrow 2Fe(OH)_2 \quad (3)$$

Other reactions can occur such as conversion of ferrous hydroxide to hydrated ferric oxide, the very familiar yellow-brown rust, by further reaction with oxygen:

$$4Fe(OH)_2 + O_2 \rightarrow 2H_2O + 2Fe_2O_3.H_2O \quad (4)$$

The rate of corrosion is dependent on the amount of available oxygen in the seawater at the steel surface. As the corrosion product (ie rust) layer builds up on the steel surface it will normally have a protective effect by limiting the access of oxygen to the steel surface, thus slowly reducing the average corrosion rate.

Depending on the conditions rust can take a variety of physical and chemical forms, differing in colour, hydration and bulk. Shielding effects of the voluminous deposit of rust may alter the course of the corrosion process. For example reaction (4) undergoes modification when the supply of oxygen is limited or restricted. Modified corrosion products arise from secondary reactions such as:

$$6Fe(OH)_2 + O_2 \rightarrow 4H_2O + 2Fe_3O_4.H_2O \quad (5)$$

$$Fe_3O_4.H_2O \rightarrow H_2O + Fe_3O_4 \quad \text{(black } \textit{Magnetite}\text{)} \quad (6)$$

The deposition of yellow-brown rust by reaction (4) frequently restricts the supply of oxygen to $Fe(OH)^2$ subsequently formed beneath it and it is common to find laminated corrosion product scales consisting of:

- an inner core of black *Magnetite*
- a thin layer of its green hydrate
- an outer coating of ordinary rust.

Under anaerobic conditions the presence of certain microorganisms such as sulfate reducing bacteria (SRB) can provide alternative cathodic reactions to support anodic dissolution of iron. Various mechanisms have been proposed but they are still the subject of debate (Lee, 1995).

SRB primarily obtain the sulfur that they need in the form of sulfate (abundant in seawater) which they reduce using organic matter as the reducing species. Some reduced sulfur is assimilated into the organism but the majority is released into the surroundings as sulfide ions (S^{2-}). These are then usually hydrolysed to free hydrogen sulfide (the sulfide ions may exist predominately as H2S, HS- or S2- depending on the pH). In this way, the SRB provide a cathodic process to support anodic dissolution of steel. The greatly simplified reactions are:

At the anode: $4Fe \rightarrow 4Fe^{2+} + 8e^-$

At the cathode: $8H_2O + 8e- \rightarrow 8H+ads + 8OH$

$8H^{+ads} + SO_4^{2-} \rightarrow 4H_2O + S^{2-}$

$4Fe^{2+} + S^{2-} + 6OH^- \rightarrow FeS + 3Fe(OH)_2$

The resultant iron sulfide (FeS) is cathodic to steel and is sufficiently conducting to replace the normal metallic component of a galvanic cell. The iron rich forms of iron sulfide which develop over relatively long periods of exposure are more cathodic to steel and more conductive, therefore corrosion rates tend to be slow at first and then accelerate with time.

A2 MARINE CORROSION

Corrosion can occur on steel either fully immersed in the seawater (called aqueous corrosion) or exposed to thin seawater film conditions (called atmospheric corrosion). A detailed description of the corrosion reactions that occur on steel in seawater is given in Appendix 1.

1.0 The marine environment

Saline waters are particularly aggressive to steel. The corrosive characteristics of seawater are primarily influenced by:

a) *Dissolved salts (ie salinity)*; the presence of dissolved inorganic salts affects both the conductivity of seawater and the solubility of oxygen in seawater. Of all the dissolved salts present in seawater, chloride is probably the most significant. This is because of its high concentration in seawater and its ability to destabilise the development of the normally semi-protective iron oxide films formed on steel surfaces.

b) *Dissolved oxygen*; the corrosion rate in unpolluted seawater is oxygen diffusion controlled and is, therefore, determined by the quantity of the oxygen supplied to the steel surface. The solubility of oxygen is dependent on temperature and the dissolved salt content of the seawater.

c) *Water velocity*; the rate of flow of seawater over a clean steel surface can increase the oxygen supply (ie enhance the cathodic reaction) and, therefore, the corrosion rate up to a limit. (Also see the sections on erosion corrosion and differential aeration below.)

d) *Temperature*; an increase in temperature results in increased diffusivity, but a reduction in the solubility of dissolved oxygen. Normal seawater temperatures vary from the 2°C to 4°C in deep oceans and up to 29°C in the tropics. Higher temperatures can occur locally, for example in the vicinity of condenser outlets. Although temperature would be expected to affect the corrosion rate in different ways, experience in service has shown that corrosion rates under fairly static seawater conditions (including tidal flow) does not increase with increasing temperature (Phull, 1997, Kirk, 1990), certainly not up to about 30°C. Marine biofouling growth (and bacterial activity) is accelerated with elevated temperatures.

e) *pH;* natural seawater is well buffered and its bulk pH normally lies between 8.1 and 8.3 (ie it is a slightly alkaline solution). However, in seawaters containing pollutants or contaminants, the bulk pH may vary somewhat, falling for example to 7.0 in stagnant basins due to the formation of hydrogen sulfide produced by anaerobic bacteria. Local pH changes can occur below marine fouling. The effect of pH within the range of 7.0 to 8.3 is of little consequence in influencing the corrosion rate of steel in seawater.

f) *Pollution*; the presence of pollutants from domestic sewage, agriculture (eg fertiliser run-off), industrial waste, wind-blown cargo dusts, etc. can make the seawater more corrosive than normal, for example by reducing the protection afforded by marine biofouling and/or increasing bacterial activity. Unprotected steel structures at sites where conditions of stagnation or restricted water movements occur, such as in estuaries and ports, are potentially more at risk to bacterial corrosion than those in the open sea (Melchers, 1994). Pollutants such as oils and greases can, however, lower the corrosion rate due to their barrier effect, reducing the rate of oxygen transfer to the steel surface.

g) *Marine organisms*; these are present in any marine environment. They may halt corrosion by forming a protective barrier at the steel surface or, conversely, they may accelerate corrosion by producing aggressive environments beneath which enhance localised corrosion.

2.0 Micro/macro corrosion cells

Corrosion is usually categorised into either general (or uniform) corrosion or localised corrosion.

General corrosion results from the formation of micro-corrosion cells on the metal surface where the distance between the anodes and cathodes is microscopic. Although metal loss only occurs at the anodic sites, the anodes and cathodes are constantly changing position resulting in a more or less even thinning over the entire steel surface.

Localised (or pitting) corrosion results from the action of macro-corrosion cells. In this case the anodes and cathodes are well separated. Unlike general corrosion, the anodic sites become permanently fixed (ie the metal loss is confined to specific areas) with the remaining surface acting as a cathode.

Localised corrosion can potentially have more significant implications on structures than general corrosion because;

- the rate of corrosion penetration at these sites can be significantly greater than that obtained during general corrosion. Local breakdown of the semi-protective rust layer or a locally damaged or degraded paint film can lead to the local formation of anodic (ie corroding) areas. The pits at such locations, once started, tend to be accelerated by corrosion currents flowing between the small anodic areas of metal at the pit and the much larger, more cathodic (non-corroding) site surrounding it, leading to an unfavourable anode-cathode area. Pitting rates, sometimes up to 20 times that for general corrosion, are quoted in the literature in the early stages of immersion, gradually tapering off over a longer period of approximately ten years exposure (AISI, 1981)
- the localised corrosion can give rise to:
  - stress raisers (ie points of stress intensification) on structural components subject to applied loads
  - holes which can lead to loss of watertight integrity of a structure, loss of backfill material and ultimately failure in the case of a retention wall.

Such macro-cells, which result in rapid rates of localised corrosion, occur because the nature of the electrolyte and/or material is not uniform over the whole of the metal/environment interface. Almost any variations in conditions of exposure at the surface of the metal can produce a corrosion current within a macro-cell leading to localised corrosion. For example, oxygen concentration cells are frequently encountered in marine service due to variations in the availability of oxygen (the cathodic reactant). This can occur either over:

- short distances (eg beneath fouling deposits or marine organisms); the area beneath the deposit becomes anodic (corroding) because it is low in oxygen compared to the surrounding metal which becomes cathodic (non-corroding)
- large distances (eg tidal versus low water zone of a steel sheet piled wall). If two areas on a electrically continuous steel surface are each in contact with electrolytes with different dissolved oxygen contents (eg due to differences in turbulence, temperature, salinity and velocity, etc), the area in contact with the

more aerated seawater (ie with the higher dissolved oxygen level) will become the cathode (non-corroding) and that surrounded by the lower oxygen level will become the anode experiencing accelerated corrosion. Such an effect can be applied to sheet piled structures in marine environments because the structure may be long and the method of installation allows for electrical continuity. Corrosion currents could then arise from both a horizontal and a vertical zonation.

Crevice corrosion is a particular example of a concentration cell formed by differential aeration. It typically occurs at mechanical joints or whenever steel is covered by an organic or inorganic layer or by marine fouling organisms, such as barnacles and weeds, which partially shield the access oxygen from the surrounding water to the steel surface.

It has been argued that "classical" crevice corrosion (such as that which occurs with stainless steel in seawater, for example) will not occur with carbon steels in seawater as evidenced by its absence at the joints (ie clutches) between steel piles.

3.0 Common forms of accelerated corrosion

In addition to pitting and crevice corrosion, the other common forms of corrosion which can produce accelerated attack on unprotected marine steel structures include the following:

a) Bimetallic (or galvanic) corrosion

When different metals or alloys are immersed in an electrolyte they establish a unique electrochemical potential with that environment. Thus different metals and alloys will have different values of potential in the same environment. If dissimilar metals are immersed in sea water and connected then the resulting difference in values of potential may produce current flow between the two electrodes. This current flow may then result in more significant corrosion at the anode and reduced corrosion at the cathode than would occur if the two metals were not connected. This corrosion is referred to as bimetallic (or galvanic) corrosion.

It is important to understand the significance of potential difference in such corrosion. Potential difference provides the driving force for current flow and defines which metal (or part of the metal) will be the anode and which the cathode and thus which metal will suffer corrosion damage (the anode) should any additional corrosion occur. However, potential difference cannot determine whether such corrosion will in fact occur nor, if it does, the rate at which it will proceed.

Many engineers may be familiar with tables of potential difference (ie galvanic series) that have, in the past, been used as a guide to predicting bimetallic corrosion and the consequences of the same. However, these tables are often misused and at best all that such tables can be usefully used for is to identify possible anodes and cathodes and the magnitude of the driving force for current flow. They cannot predict what will happen nor the rate at which corrosion might occur. It is also incorrect to state that the rate of corrosion is related to the magnitude of potential difference between the metals. The folly of such an approach can be illustrated by the case of PWC that might occur in the HAZ of welds immersed in seawater. Typically, the potential difference may be of the order of 20 to 50 mV which is a relatively small difference. However, under certain conditions corrosion rates of the order of 1 to 2 mm/year maybe experienced.

Whether bimetallic corrosion occurs is influenced and determined by factors that are not directly related to potential difference and include:

- the presence of films or scales on the surface (particularly on the cathode) that may delay the reduction of oxygen (the normal cathodic reaction in aerated seawater)
- the relative areas of anode and cathode, as this determines the current density (amps/unit area) at both the anode and cathode. Small anodes connected to large cathodes can sustain high rates of corrosion because the current density at the anode will be high. This is the primary reason for the high rates of damage that may occur in PWC as the HAZ (ie the anode) has a small area compared with the infinitely massive cathode that is the remaining steel connected to the anode
- the geometric shape of the electrodes and the conductivity of the surrounding electrolytic solution as these determine the resistance of the electrode to the environment and the ease with which current may flow.

It follows from the above that problems of bimetallic corrosion can be minimised by taking some simple precautions:

- avoid using dissimilar metals when immersed or buried in electrolytes unless there is documented evidence that the particular couple in that environment will not cause a problem. Guidance on this may be obtained from BS PD6484
- avoid connecting dissimilar metals in such a manner as to create large differences in surface areas between the anode and cathode or try to create situations in which the possible anode is of greater area than the associated cathode
- take particular care with coated surfaces, small areas of damage to coatings on an anode may result in intense local corrosion at the area of coating damage
- care is needed at welded connections to minimise/eliminate the risk of bimetallic corrosion through the use of an appropriate welding consumable (ideally this should be cathodic to the steel parts being joined) and a qualified welding procedure. For critical elements the welding procedure qualification tests should include an assessment of bimetallic corrosion risks.

It should be self evident that the risk of bimetallic corrosion is greater where a metal is permanently immersed than in conditions of intermittent immersion or atmospheric exposure and clearly it will not happen if the couple does not become wet. However, the risk of bimetallic corrosion in conditions of marine exposure should not be underestimated and value guidance on this can be obtained from BS PD 6484.

In addition to the obvious risks of bimetallic corrosion arising from connecting different metals and alloys in an electrolyte there are other less obvious situations where bimetallic corrosion can occur that involve connection of a metal to non-metallic conductor or semi conductor. Such couples include:

- connection of metals to elemental carbon as might occur with piles driven into ground containing ash, clinker or coal tailings. In this case the metal may well suffer significant additional corrosion
- connection to conductive elastomers such as those compounded with excessively high levels of carbon black commonly used in these materials as a UV stabiliser
- the presence of specific scales or films on the surface of a metal. The best known examples in this category are the presence of mill scale and iron sulfide films on the surface of steel and these are discussed below as they are particularly relevant to seawater corrosion and ALWC.

Mill scale is a high temperature oxide formed during steel manufacture as a result of the hot rolling process. Unless the steel is weathered for sufficient time or abrasively blast cleaned prior to installation the mill scale will remain in service. Mill scale is strongly cathodic to the underlying steel and at breaks in the scale, which will occur as the steel weathers or the surface is damaged, high rates of localised corrosion may occur if the area of mill scale is very much greater than the exposed steel surface. It can lead to rapid pitting giving pit depths up to 1.25 mm on as-rolled steel after six months immersion in seawater (L L Shrier, "*Corrosion*", Butterworth-Heinemann Ltd). Steel exposed with mill scale present will be pitted about three times as deeply as descaled steels for such short periods of exposure as a few months (F L LaQue "*Marine Corrosion*" John Wiley & Sons, 1975). In most cases the effect is relatively short lived (6 to 12 months) as the mill scale itself gradually detaches from the steel surface.

Iron sulfide films may form on the surface of steels in service as a result of SRB activity. Again these films are strongly cathodic to the underlying steel surface, and provided they remain intact, they will not cause problems. However, such films are not particularly adherent and can easily be damaged. At such areas of damage the exposed steel will be anodic to the surrounding film that can act as an efficient cathode and sustain high local rates of corrosion.

b) Flow assisted corrosion mechanisms

Any mechanism that enhances the supply of oxygen (the cathodic reactant) to the bare steel surface immersed in seawater would be expected to increase the corrosion rate. Examples include thinning or removal of corrosion products or protective layers of marine biofouling growth that would otherwise stifle further rusting.

Erosion corrosion occurs when the velocity of seawater over the metal surface is sufficiently fast to strip away or continually remove otherwise semi-protective surface films. This allows the corrosion to continue unchecked at a much more rapid rate because the metal substrate is now in direct contact with the corrosive seawater. Marine organisms or fouling can become attached at times when the water is moving at low velocities and can serve as a barrier to reduce or eliminate the accelerating effect of higher water velocities (Inco, 1973). However, even at relatively low velocities abrasion and/or corrosion can be significant where the water includes abrasive particles such as entrained mud or sand, for example sand scour with tides and/or current movements. Such mechanical abrasion can also be caused by propeller scour during ship berthing (PIANC, 2004) particularly bow thrusters where water is perpendicular to the quay.

c) Stray currents

Whenever external DC (direct current) electrical sources are situated near marine construction, stray currents from non-insulated regions may flow to and from these metallic structures. Where the stray current enters the structure, the electrical potential of the structure will be lowered and corrosion rates reduced. However, where the stray current leaves the structure, the opposite is true and high rates of local metal loss can occur. Direct current from improperly grounding welding generators, ship service systems, nearby cathodic protection systems and electric railway systems are potential sources of stray current corrosion damage. Similar effects, albeit short term, may also occur when magnetic objects pass through electrical fields eg dock cranes through fields around cables. It is essential that stray currents are eliminated at their source or that the structure be suitably grounded (AISI, 1981).

d) Microbially induced corrosion (MIC)

Also sometimes referred to as "*Microbiologically influenced corrosion*" or "*Microbially assisted corrosion*". The microbiological potential of a marine environment is immense and micro-organisms are well known to contribute to corrosion processes in such environments. They can often make a significant contribution to an overall corrosion process without being solely responsible (Stott, 1998). For example, they can:

- produce high concentrations of aggressive by-products (eg acid production, hydrogen sulfide, etc)
- produce localised changes in oxygen concentration
- lead to the formation of differential aeration cells or crevice corrosion due to their physical presence as a slime or poultice on the surface
- it has been suggested that they can directly stimulate cathodic reactions on steel surfaces, but there is not much real evidence for this.

The types of micro-organisms contributing to corrosion comprise (Tiller, 1983):

- aerobic bacteria (ie only grow under oxygenated conditions)
- anaerobic bacteria (ie only grow in the absence of oxygen).

The most important group of micro-organisms associated with the corrosion of steel in marine environments is the sulfate reducing bacteria (SRB). They have been responsible for rapid and severe localised corrosion on steel in marine environments with reports of localised (pitting) corrosion rates of up to 8 mm/year in the case of seawater ballast tanks (Campbell, 1990). Details on the mechanism by which SRB are thought to contribute to the corrosion of carbon steel are given in A1. SRB are anaerobic bacteria and reduce inorganic sulfate (abundant in seawater) to sulfide, which appears as hydrogen sulfide (gaseous or dissolved), bisulfide (HS-) or sulfide (S^{2-}) ions or metal sulfides or a combination of these depending on conditions. They are most likely to be found where fluid velocities are low or in stagnant areas or under scale deposits/debris/sludge. Corrosion of steel by SRB is characterised by the formation of black iron sulfide and shiny pitting on the steel surface. The iron rich forms of iron sulfide compounds which develop on the steel surface over relatively long periods of exposure are thought to be more cathodic to steel and, therefore, corrosion rates tend to be slow at first and then accelerate with time (Stott, 1998). Although SRB are strict anaerobes they will tolerate long periods in well aerated environments such as seawater where they will exist in a dormant form until suitable conditions allow them to grow. The normal constraints on SRB numbers are sulfate availability, presence of toxic substances (eg oxygen) and unsuitable temperatures or pH ranges. SRB (*Desulfovibrio*) will grow in pH range 5 to 10 and at temperatures between 5 and 50°C.

Oxygen consumption by the activity of other bacterial colonies such as aerobic sulfur oxidising bacteria (SOB) and acid producing bacteria (APB) species present initially on the steel surface can provide the necessary oxygen-free conditions beneath them to support the growth of SRB. Consequently, SRB are often found operating in close association with SOB and APB at the base of even thin layers, supporting each other's growth. When the components that promote microbial growth form a colony on a surface they create what is referred to as a biofilm.

SOB oxidise reduced sulfur compounds (eg soluble sulfide) to sulfur or sulfate and are involved with SRB in what is referred to as the sulfur cycle. Thiobacillus (an aerobic SOB) obtains its energy by oxidation of inorganic sulfur compounds (including sulfides produced by SRB) to sulfuric acid. If a high concentration occurs locally their activity may be great and result in a zone of pH2 or lower. Some species of SOB (eg the anaerobic green and purple sulfur bacteria) as well as Algae rely on exposure to light (UV) to provide their energy.

APB are slime forming bacteria and are more common in waters of low salinity (ie brackish). They are able to degrade hydrocarbons (eg oil) to produce short chain volatile fatty acids (VFA). SRB consume VFA as a carbon source and produce acetic acid and carbon dioxide in addition to hydrogen sulfide.

Positive indications of SRB activity are:

- the smell of hydrogen sulfide (which resembles bad eggs) particularly when the black deposit is contacted by acid. The liberation of hydrogen sulfide on exposure to acid is a simple test that can be used to distinguish black iron sulfide (formed as a result of SRB activity) from black *Magnetite* (iron oxide formed under limited oxygen conditions). Care should be taken not to inhale hydrogen sulfide as it is toxic
- the presence of a black coloured deposit of iron sulfide on the steel surface. Identification of the iron sulfide compounds as *Mackinawite, Smythite* and *Greigite* in the deposit (Baldwin, 1998; McNeal 1990). Note that iron sulfides can easily oxidise on exposure to oxygen (air), the black colour disappearing within hours. The iron sulfides typically oxidise to hydrated iron oxide (ie brown rust) and elemental sulfur (yellow) or sulfates
- the presence of smooth, shiny pitting described:

    a) In the case of pipeline steels (Baldwin, 1998) as "Normally a larger pit will have smaller individual pits inside it. In more developed cases the pits may overlap each other. Striations and tunnelling parallel to the steel surface are also pointers to MIC, but it can also occur without these signs. Sometimes the pits are covered by a nodule (tubercle) of material composed of the microbes and the material they consume and deposit. As pits deepen and become more numerous, individual pitting becomes less obvious".

    b) Inside ballast tanks (Campbell, 1990) as "The pits were large being typically 8-35mm across at the point of maximum width and 1.5-9 mm in depth. The majority of pits were rounded in shape with gently sloping sides. In some regions, blackened material was concentrated near the bottom of some pits. Removal of surface scale revealed darkened corrosion products".

- positive tests confirming the presence of SRB. They can be identified by API method RP-38 or serial dilution methods using marine media. Several test kits and sessile probes are also available for SRB and other types of bacteria
- encrustations or "tubercules" on the metal may also be associated with the anaerobic environment required by SRB.

4.0 Traditional corrosion behaviour

This section describes the "traditional" (ie oxygen diffusion controlled) corrosion behaviour of unprotected marine steel structures in unpolluted seawater subject to tidal influences.

Chapter 2, Figure 2.2 illustrates the widely accepted typical corrosion profile for unprotected steel in unpolluted seawater containing dissolved oxygen and represents the varying conditions of immersion which unprotected steel may encounter when used in marine applications. The average corrosion rate values are as a result of measurements taken in several worldwide studies carried out over periods of up to 50 years. Much of this information has been obtained from test panels or by measurements on pier supports or piling in marine environments and is recorded in several books and published papers. In practice, some variation in actual corrosion rates would be expected from installation to installation and, although the pattern of corrosion will often remain the same, the curve may be displaced to either the left or right depending on local conditions. These empirically derived values of corrosion rate assume unpolluted seawater. Some variation in excess of the corrosion rate values quoted would also be possible on unprotected, new steel during the initial stages of exposure (ie prior to the build up of a semi-protective scale, marine organisms).

Examination of this corrosion profile for port steel structures reveals several distinct areas of attack and it is convenient to divide the marine pile into six zones each having a characteristic corrosion rate (BS6349-1: 2000). The corrosion mechanisms operating in these zones of marine structures have been well documented in the literature, some zones being more vulnerable to attack than others.

- **atmospheric zone;** *the area between splash zone and top of the structure*. This area is affected mainly by seawater spray. In order for corrosion to occur at this zone it requires the presence of water present either as liquid (ie a thin film of moisture) or vapour (ie water pick up by hygroscopic salts such as chlorides present on a metal surface). The corrosion in this zone will be controlled to a large extent by factors such as:
  - the duration and frequency of moisture film contact
  - atmospheric humidity
  - the rate at which the surface dries
  - the build up of aggressive sea-salts (eg chlorides) onto the steel surface
  - the level of pollutant gases and contaminants present in the surrounding atmosphere
  - the degree of exposure to wind and rain.

  Overall, the corrosion rates are far lower than those evident at the splash zone due to the comparatively less aggressive environment in this zone.

- **splash zone**; *the area from mean high water (MHW) up to a height depending on mean wave height*. In the zone just above high water level, the steel surface is subject to splashing and the rust films have little opportunity to dry out and are less protective. The corrosive conditions are aggravated by the high oxygen content of the thin film of seawater on the steel surface. This zone represents the most severe conditions and is considered the zone where the highest loss in metal thickness (ie highest uniform and localised corrosion rate) would be expected to occur. Actual corrosion rates will depend on exposure of the structure to wind and wave action ie sheltered harbours would be less affected
- **tidal zone;** *the area between MLWS and MHW*. This zone is repeatedly immersed in seawater and then exposed to air by the ebb and flow of the tide. Potentially the tidal zone represents a severe corrosive environment but in spite of this, the metal wastage in this zone is lower than in other zones. A particularly low loss area exists in the tidal zone about halfway between MHWS and MLWS. The explanation for this (Humble, 1949) is that the well-aerated areas in the tidal zone become strongly cathodic while the metal just below low water level becomes anodic. Steel surfaces in the tidal zone, are exposed to highly aerated seawater as the tide flows compared with the lower oxygen content of the relatively stagnant seawater around the submerged zone surfaces forming a macro-oxygen concentration cell (OCDI, 1998). At some sites floating oil deposits can coat the steel surface in the tidal zone and this has been considered to provide partial protection by filming the metal surfaces (AISI, 1981). The effect of biofouling on general corrosion in the tidal zone is believed to be beneficial by acting as an oxygen diffusion barrier. However, it has been suggested that biofouling can produce enhanced localised corrosion due to

acidic conditions generated by the biochemical breakdown of dead organisms or by local shielding of surfaces (ie differential aeration).

This zone may be absent or reduced in an impounded dock system where there is no tidal variation and variations in water level are insignificant

- **intertidal low water zone (ILW) zone;** *the area between a point approximately 0.5 m below LAT and MLWS*. A significant metal loss can occur a short distance below MLWS attributed to the action of a macro-cell as described above. Another explanation sometimes given is that the concentration cell may be set up as a result of oxygen producing fouling.

  This zone is the location where a higher corrosion rate can occur than in the other submerged zones. Rates similar to those seen in the splash zone can occur in some locations although the corrosion rates in this zone can vary. Localised corrosion can also occur, for example, if this area is covered by oxygen shielding marine organisms or surface rust and/or marine organisms are repeatedly removed. Marine biofouling can also provide a suitable environment for colonisation by micro-organisms capable of affecting the corrosion process

- **continuous immersion zone;** *the area from mud line (sea bottom) up to the lower limit of the ILW zone (note that this zone does not exist in those locations where the mud line is above the MLWS)*. In areas that are continually submerged, attack on the surface is governed principally by the rate of diffusion of dissolved oxygen in the seawater through layers of rust (scale) and marine organisms usually present on the steel surface, both of which limit corrosion. Higher rates can be expected if the rust layer and marine growth are not present. A secondary area of accelerated corrosion may occur in some locations at or near the mud line due to water velocity and the resultant scouring action by abrasive sand repeatedly removing corrosion products and marine organisms from the steel surface (ie erosion corrosion)

- **mud (embedded) zone;** *the area of total burial in mud or soil*. Other than at locations where the mud is seriously contaminated with pollutants or contains high levels of anaerobic bacteria (SRB) which attack the buried steel (AISI 1981), the area below the mud line is often subject to minimal corrosion in line with that found in the submerged zone. This is due to several factors including the lower availability of dissolved oxygen, the lower conductivity of mud compared with seawater and the undisturbed rust films formed on the surface of the steel.

A3 FINDINGS OF PREVIOUS RESEARCH STUDIES ON ALWC

Various research studies (eg Moulin, 1998) have been carried out primarily funded by the steel manufacturers. Japanese researchers have also studied low water corrosion and have published their findings (eg OCDI, 1998; Fukute, 1990; Tsuchida, 1985).

1.0 The findings of a joint research project involving a survey of five west European countries carried out over a three year period by four steel makers and TESPA (Technical European Sheet Piling Association) were reported in 1995 (Moulin, 1995). Analysis of the reported characteristic black liquid layer next to the steel surface at low water level showed that it contained high levels of iron sulfides and, relative to the seawater, the pH was found to be acidic and contained significant levels of ferrous, sulfate and chloride ions. It was suggested in this work that the outer layer, which is covered by living organisms, probably acts as a barrier to the diffusion of oxygen. Analysis of 36 samples removed from problem sites showed high contents of:

- *magnetite* (10 to 50 per cent) – which is a normal rust constituent formed under limited oxygen conditions
- "*green rust 2*" (30 to 80 per cent) – an unstable iron oxy-hydroxide sulfate complex which has been linked to corrosion by SRB
- *iron sulfides* (5 to 20 per cent) – probably produced as a result of sulfate reducing bacteria (SRB) activity.

It was reported that the composition of the SRB population was different between areas affected with the problem and those without the problem and that to produce such high corrosion rates, the SRB needed other bacteria to be present, such as aerobic bacteria.

2.0 In another study (Beech, 1996), samples of biofilms recovered from the surfaces of reportedly affected and unaffected piling in a south coast UK harbour were analysed. The biofilms present at low water level were found to contain a diverse range of aerobic and anaerobic bacteria known to play a role in the marine corrosion of steel. As previously, the composition of the microbial species varied depending on the region from which the samples were taken. In general, higher levels (3×10^7 cells/cm^2) of all types of bacteria were detected in the biofilms recovered from affected areas of the piling (with SRB being the dominant species) compared with biofilms removed from non-corroding sites (3×10^4 cells/cm^2). X-ray diffraction (XRD) analysis on biofilms removed from low water level at the problem sites revealed largely iron sulfides and from non-problem sites the presence of "*Green Rust 2*" and *Magnetite*.

3.0 The European Coal and Steel Communities (ECSC) joint research project investigating the problem reported the findings from the first phase of this work in 1997 (Johnson, 1997) and published the results of the second phase of this research in 2001.

The ECSC first report gives a detailed description of the problem together with an investigation into the corrosion mechanisms. Electrochemical potential measurements were made on existing pile structures and showed that significant macro-cathodic and macro-anodic areas are initially formed in the tidal zone and at the low water level respectively. It was reported that these differential conditions did not persist after about

200 days (presumably due to the build up of corrosion products and marine fouling on the surface of the steel and/or loss of mill scale). It was concluded, therefore, that macro-electrochemical effects do not provide a viable explanation of the problem. Also the pattern of holing on different pile geometries could not be explained by any changes in electrical current due to either orientation or shielding effects.

Around holes, striations in the form of thin slivers of steel and tunnelling, parallel to the steel surface, were also observed.

Surface deposits at high water level (HWL) were found to be well adhered and comprised a single stratum of red-brown corrosion product covered by a thin layer of marine fouling organisms (mainly barnacles and weeds). In contrast, the fouling layer at the low water level was found to be thick, non-uniform and poorly adherent on both ALWC affected and unaffected piles. The detached layer exhibited clear stratification and was reported to consist of:

1. An outer layer of calcareous deposits together with living shellfish, algae, weed, etc in direct contact with the moving seawater.
2. An intermediate layer which was non-uniform and highly porous consisting of red-brown corrosion products which included seawater derived compounds, oxygen, bacteria and so on.
3. An inner layer in contact with the steel which comprised a black sludge film only formed at the low water level, this black film varied in nature being present as a transparent liquid film in which black particles are in suspension, or a semi-liquid sludge layer or even a wet, but firm, solid layer; exposure of the black film to the atmosphere led to rapid oxidation and the formation of red corrosion products. Analysis of the black liquid sludge film showed it contains high levels of iron sulfide, probably in mildly acidic solution (approximately pH6), and "*Green Rust 2*".
4. An underlying steel surface that was bright, clean and very heavily pitted.

Significantly, all these analyses of surface deposits showed no clear differences between problem sites (referred to in the report as "*ALWC affected*") and those not (yet?) suffering with the problem (referred to as "*non ALWC sites*"), although clear differences between the surface deposits at the low water level and in the tidal zone were found. Also at low water level no clear differences between surface deposits present on outpans (ie at holes) and the inpans (ie no holing) were found. Furthermore, these observations were common to both problem and unaffected sites.

In Boulogne, large "blisters" were reported to have formed on the steel piles. In a 1978 survey at a UK port (Morely, 1978), blistering was also reported to have been observed. Analysis of liquid removed from within the blisters at Boulogne gave the following results:

1. The liquid phase showed high chloride (twice that of seawater) and ferrous concentrations and a reported pH of about 3.
2. The black solid suspension showed the presence of *Magnetite, Goethite, Lepidocrocite, Maghemite*; iron sulfide and "*Green Rust 2*".

The ECSC second report gives further information on the nature of the surface products and condition of the steel surface itself at the low water level and provides the results of additional analyses on surface deposits and seawater from 22 sheet piled structures at five different geographical locations within Europe.

The report refers to "*ALWC*" and "*NLWC*" (normal low water corrosion) structures. However, *NLWC* did not mean structures exhibiting normal hydrated iron oxide scale and marine deposits on the steel surfaces in the low water level zone (ie as for the ALWC structures, biofilms were also found to be present at these sites).

An *ALWC* structure was defined as one:

- that exhibits holes at the low water level
- that has a high local corrosion rate at this level (at least 0.3 mm/year)
- that shows heavily pitted, bright steel surface around holes when the corrosion product are removed
- where removal of the corrosion products is easy because they are poorly adherent to steel.

An *NLWC* structure was defined as one:

- that does not exhibit a corrosion hole at the low water level
- where there is no local thinning at the low water level
- the mean corrosion rate at this level is about 0.1 mm/year.
- that shows a slightly pitted, dull steel surface when corrosion products are removed at the low water level
- where corrosion product removal is difficult because they are very adherent to the steel.

Site tests were performed to investigate the anode/cathode ratio effects arising from the macro-oxygen concentration cell formed between the tidal zone and low water level zone. These tests were done in an attempt to explain why the problem occurs with apparently similar corrosion characteristics at locations with widely differing tidal ranges. It was concluded that where strongly anodic (ie corroding) areas develop in the low water level zone the surrounding cathodic areas within the low water level zone itself may well be an important factor rather than just the cathodic tidal zone area. It was suggested that this could explain why the problem also occurs at locations where the tidal range is small.

As before, the surface deposits at low water level were stratified and the three main layers were identified. The chemical characterisation of these surface deposits did not reveal any major differences between *ALWC* and *NLWC* structures. However, microbial and chemical analysis did indicate higher levels of the following in low water level deposits at *ALWC* versus *NLWC* sites:

- SOB
- algae
- calcium and chloride
- organic matter
- thickness and porosity of corrosion products were higher.

and lower levels of:

- pH beneath corrosion products
- invertebrates (eg mussels, barnacles)
- phosphorous compounds (note that phosphorus, in the form of phosphates, is required for the enzymes responsible for sulfate reduction by SRB)

- heavy metals (eg zinc, chromium)
- manganese and total sulfur.

No major differences were found in the distribution of iron, oxygen, calcium, sulfur and chlorine (using EDAX elemental mapping) between the surface deposits taken from the inpans and outpans.

X-ray diffraction (XRD) on samples of total corrosion products (ie the combined three layers), but exposed to air, taken from *ALWC* and *NLWC* structures revealed the presence of iron sulfide, ie *Troilite* and *Pyrite* forms, normal constituents of rust and elemental sulfur (indicative that oxidation of the iron sulfides had occurred to some extent).

X-ray photon electron spectroscopy (XPS) analysis on the inner layer black material (kept away from air) indicated that iron sulfides and ferrous ion (Fe^{2+}) compounds were present at high levels even at *NLWC* sites.

Overall the analytical data showed that the corrosion products (mainly iron) at ALWC areas were associated with higher levels of sulfur and sulfate as well as increased microbial and chemical activity.

A4 RISK ASSESSMENT MODEL FOR ALWC

An attempt was made as part of the (ECSC) joint research project to develop methods of assessing the risk of ALWC to any specific site using statistical analysis (Gubner, 1998 and 1999). Similar data for the statistical analysis was obtained from various harbours throughout Europe using the following parameters:

Site parameters

- mean tidal range
- presence of invertebrates
- presence of algae.

Corrosion product parameters

- thickness of corrosion products
- pH underneath corrosion products
- MPNSOB (most probable number method)
- organic carbon
- organic hydrogen
- organic nitrogen.

Seawater parameters

- redox potential
- total organic carbon.

It was concluded that:

- some of the parameters can be easily measured and used for the monitoring of an existing structure
- measurement of the three parameters would allow a direct comparison with the measurements obtained from other existing piling structures which could provide some indication of risk
- the risk assessment model can also be applied to new structures but with a lower level of reliability
- ALWC may be the result of a number of factors rather than triggered off by one key factor.

A5 THE CIRIA QUESTIONNAIRE: Survey on Accelerated Low Water Corrosion

This questionnaire was developed from a survey conducted by the ICE Maritime Board. It has been created as part of a research project by CIRIA, the leading independent provider of construction research and innovation. It is of particular relevance to owners, operators, designers, contractors and other parties dealing with marine structures. The project will produce a comprehensive and user friendly report, to be published in summer 2005, addressing all aspects of ALWC – what it is, how to recognise it, how to treat it and how to prevent it.

WHAT'S IT ALL ABOUT?

"Biocorrosion threat to UK coastal defence"

In some ports and harbours there are particular conditions which can cause localised, accelerated loss of metal on unprotected steel structures. This rapid form of attack, which results in extensive and costly remedial work at an unexpectedly early stage in the life of a structure, has been called ALWC (Accelerated Low Water Corrosion).

We are carrying out a fact finding survey on this high profile problem and would ask that you contribute by completing the attached questionnaire. The main objectives of this survey are to:

- identify the incidences of ALWC in the UK and abroad
- determine what has been done in the past to mitigate the ALWC problem and obtain feedback as to whether such measures worked.

WHY SHOULD I PARTICIPATE IN THIS SURVEY?

- **owners and operators**

Do I have an ALWC problem and, if so, what are the implications and expected costs?

As the owner of marine infrastructure you have a legal obligation to provide safe facilities. In order to avoid unexpected repairs and associated costs it is essential that you identify the problem early. Your participation in this questionnaire will help us to establish the extent of the problem and to understand your concern and decisions behind the measures that you may have taken to deal with it, either on new build or existing structures. Information on how effective the chosen preventative remedial options have been will also be of significant value. This information will allow practical guidance, based on existing knowledge, to be provided in the final CIRIA report. The intention is to save you money in the long term!

- **contractors/maintainers**

What is ALWC and what needs to be done to deal with the problem?

As a contractor or maintainer you may have already heard of the ALWC problem and recognise its rapid deteriorating effects on structures. Perhaps you already have first hand knowledge of dealing with ALWC. By providing information to this survey on the sort of ALWC problems you have encountered and how you have dealt with them, you will enable us to pool your experiences on the wide range of possible inspection, repair and maintenance techniques available and allow us to advise on their suitability for particular circumstances.

- **engineers/designers**

Why does ALWC occur and where can it be expected?

You have designed/inspected/maintained maritime structures where ALWC is anticipated or present. You are acutely aware of the lack of any definitive guidance on how to deal with the ALWC problem and are developing solutions in isolation. Your participation in this questionnaire will enable us to pool information on the similar measures that can be taken to deal with the ALWC problem on new build structures and the repair/protection/ monitoring options on existing structures. The latter covers both, those structures in the early to intermediate stage of ALWC attack and on holed structures. The information you provide will help us to produce an invaluable guidance document within the final CIRIA report.

HOW CAN I IDENTIFY ALWC?

A typical description of a first encounter with the ALWC problem is given below. Unfortunately, in situations where facilities have not been inspected the first indication of the problem is somewhat different in that someone notices by chance a hole in a structure at low water level (LWL), or subsidence at the rear of the structure. Areas affected by ALWC corrosion are identifiable by their characteristic appearance:

1. A soft, spongy, bright orange/brown outer layer.
2. A thinner, gelatinous, loosely adherent inner black liquid/paste layer beneath which;
3. The steel is bright, clean and shiny with extensive pitting.

"With my hand, I wiped away a vivid orange discoloration from the pile surface and observed a film of black sludge beneath, revealing a bright shiny, dished and pitted metal surface below".

Close up of ALWC (left): Note the vivid orange colour on the surface, with a black, slimy layer underneath covering clean, bright steel.

WHERE SHOULD I LOOK FOR ALWC?

ALWC is typically found in a narrow band at or around LAT. It is not uncommon for some structural steel piles (SSP's) in the susceptible low water zone to be affected by ALWC corrosion while others in close proximity appear to be immune.

Above LWL little metal wastage may be evident. In order to ensure that the surveyor is not misled into believing that this is typical for that structure, it is essential that any inspection is carried out at one of the very lowest spring tides in order to recognise the ALWC problem. Note that in some cases the problem areas may extend down to seabed level and would not be seen without a diving survey.

ALWC on U Piles (round holes; centre of outpans)	ALWC on Z Piles (elongated slits; corner of outpans)

A feature of the ALWC problem is that there is a consistent and specific difference in geometry of damage between Z-profile (ie Frodingham) and U-profile (Larssen) piles. The former suffer from elongated slits in the corner of the outpans, whereas the latter exhibit round holes in the centre of the outpans. ALWC can also be found on other unprotected steel structures and pile geometries eg tubular, H-piles, and so on. The person completing this questionnaire, may have observed ALWC in areas or in forms and shapes that have not yet been identified due to the lack of correlated information.

QUESTIONNAIRE

Section 1 DETAILS

EXPLANATORY NOTES

We realise that filling in a questionnaire can be tedious and time consuming task, however we have tried to made it quick and simple! You can fill in most of the questionnaire by ticking the appropriate box. If more than one answer applies tick all the relevant boxes. If you have more than one structure, please fill in a questionnaire for each structure.

Please feel free to print out the questionnaire and add any comments to any section. A return address is provided at the end of this questionnaire. However, you are also welcome to type in any comments, save the changes and return this questionnaire to us by e-mail.

CONFIDENTIALITY AGREEMENT

I require that the information provided in this questionnaire is kept confidential; this applies in particular to personal and company details and the exact location. No, I do not require that the information is treated confidentially	☐
I permit all other information such as data on corrosion progression and the region (100mile radius) to be identified and used for further analysis. No, I do not want any reference made to the location	☐

DETAILS

Name	
Company	
Address	
Contact telephone number	
Contact fax number	
e-mail	
Company type	Owner ☐ Consultant ☐ Contractor ☐ Other ☐

Section 2 STRUCTURE DETAILS

2.1 Installation type

Port	☐
Marina	☐
Coastal Defence	☐
Other	☐

2.2 Age of installation

2000	☐
1995	☐
1990	☐
1980s	☐
1970s	☐
1960s	☐
1950s	☐
not known	☐

2.3 Type of structure and construction element

- retaining or sea wall

Sheet	Z (Frodingham)	☐
	U (Larssen)	☐
Cellular construction (Straight Web Frodingham)		☐

- Composite Wall System (Combiwall)

Master/King Piles	H	☐
	Tube	☐
	Box	☐
Intermediate Piles	U	☐
	Z	☐

- Isolated bearing piles supporting

suspended structures ☐

Type	ie Rendhex	☐	Larssen Box	☐	tube	☐
	H box	☐				

- Fenders

Type, Please describe ______________________________

- others
  - ◆ Dock gates ☐
  - ◆ Non structural fixtures ☐

Please describe ____________________________

- the structure is facing

  North ☐ South ☐ West ☐ East ☐

- length of structure

<50 m	☐	<250 m	☐	<1.0 km	☐
<100 m	☐	<500 m	☐	>1.0 km	☐

- protection from corrosion

  the structure was protected from corrosion with built-in measures. The protection measures were:

  don't know ☐
  paint coating ☐ Colour ____________________
  coal tar epoxy ☐
  bituminous ☐

  sprayed metal coating (eg aluminium) ☐

  galvanising ☐

  other coating (eg chloroprene rubber) ☐

  wrapping

  petrolatum (eg Densotape) ☐
  plastic ☐
  grout/concrete jackets ☐

  corrosion allowance ☐

  cathodic protection

  galvanic anodes ☐

  impressed current ☐

  others ____________________(<describe)

  none ☐

- type of steel
  - ◆ don't know ☐
  - ◆ mild steel ☐
  - ◆ high Yield ☐
  - ◆ weathering steel ☐
  - ◆ cast (iron) ☐
  - ◆ cast (steel) ☐
- depth of water at low water
  - ◆ dries out ☐
  - ◆ 1m ☐
  - ◆ 2m ☐
  - ◆ 4m ☐

- 6m ☐
- 10m ☐
- 12m ☐
- 14m+ ☐

- type of cargo
  - fish ☐
  - animal feed ☐
  - timber ☐
  - containers ☐
  - other ☐
  - bulk ☐
  - sand/aggregates ☐
  - coal ☐
  - soya ☐
  - grain ☐
  - other ________________

2.4 Environment

Type of water		**Type of exposure**		**Water current**	
sea water	☐	river	☐	none	☐
fresh water	☐	estuary	☐	weak	☐
brackish	☐	open Sea	☐	medium	☐
		lake	☐	strong	☐
		dock	☐		

Water pollution		**Type of pollution**		**Bed material**	
low	☐	oil	☐	sand	☐
medium	☐	sewage	☐	silt	☐
high	☐	chemicals	☐	gravel	☐
		other	☐	chalk	☐
		none	☐	others:	
				specify:	

Water temperature (°C), please tick box for minimum and maximum temperature range:

0-3	☐	10-12	☐	19-21	☐	28-30	☐
4-6	☐	13-15	☐	22-24	☐		
7-9	☐	16-18	☐	25-27	☐		

Abrasive wear

(eg ship mooring, erosion due to sand slurry, silt, etc)

none ☐
low ☐
medium ☐
high ☐

General observations

Marine fouling present on structure (eg barnacles) ☐
Orange/brown/black corrosion products present on structure ☐
Holes present ☐
Blisters present (concealing ALWC sites) ☐
Incidence of ALWC is different between shaded and sunny locations ☐
Incidence of ALWC is a function of sea water turbidity? ☐

Type of fenders

- none ☐
- floating ☐
- timber ☐
- other ________________

2.5 Inspections and asset management

The structure is subject to a defined asset management regime ☐

The structure is managed and maintained on an "as needs" basis ☐

Type of routine inspections and inspection intervals

	every year	2 years	3 years	4 years	5 years	> 5 years
Visual from surface	☐	☐	☐	☐	☐	☐
Visual (boat)	☐	☐	☐	☐	☐	☐
Visual (divers)	☐	☐	☐	☐	☐	☐
Other testing	☐	☐	☐	☐	☐	☐
For a combination, please click the relevant boxes						
Was any other testing used? Please specify.						
The structure is not inspected at regular intervals			☐			

2.6 Inspection records

A database of inspection records is kept ☐

Starting from (date: mm/year) ______________________________

No inspection records are kept ☐

A computerised database of inspection records is kept ☐

Inspection records can be reviewed from

- date of construction ☐
- over past 40 years ☐
- over past 30 years ☐
- over past 20 years ☐
- over past 10 years ☐

2.7 Corrosion observations

No significant corrosion has been observed ☐

Significant corrosion has occurred on the structure ☐

The structure has been inspected for ALWC ☐

ALWC corrosion has been observed ☐

Corrosion has occurred, but I am unsure if it is ALWC ☐

Corrosion is located

- just above mean low water springs 0
- at mean low water springs 0
- just below mean low-water springs 0
- elsewhere ____________________ (describe)

Details of corrosion at low water level

Corrosion has first been observed in _______ (type in the month and year)

Since then the thickness of the steel has been reduced by a maximum of

1 mm	☐	10-12 mm	☐
2-4 mm	☐	12-14 mm	☐
4-6 mm	☐	14-16 mm	☐
6-8 mm	☐	16-18 mm	☐
8-10 mm	☐	> 18 mm	☐
		corroded through	☐

If the sections have corroded through, they appear to look round ☐

slits ☐

both ☐

other forms ☐

The location of corrosion is on

outpan	☐	outpan and inpan	☐
inpan	☐	outpan or inpan	☐
webs	☐		

The distribution of corrosion appears to be

random ☐

in groups ☐

in groups at regular intervals ☐

The following possible explanations have been proposed for when the corrosion occurs in groups. Please tick the ones that could apply in your case.

position of berthing vessels ☐

frequency of berthing ☐

berthing of cathodically protected vessels ☐

use of bow-thrusters ☐

proximity of other cathodically protected structures ☐

use of floating fenders ☐

entrances of drains, sewers, etc ☐

unloading of potentially corrosive cargoes ☐

other reasons ______________________________ (describe)

2.8 Repair section

The structure has been repaired due to corrosion using one of the following measures:

This was the	1st	2nd	3rd	>3rd time of repair*
Paint coating	☐	☐	☐	☐
Specialist Anti-				

ALWC Coating	☐	☐	☐	☐

Which one? ______________________________

metal spray coating (eg aluminium)	☐	☐	☐	☐
other coating (eg chloroprene rubber)	☐	☐	☐	☐
wrapping				
petrolatum	☐	☐	☐	☐
plastic	☐	☐	☐	☐
grout/concrete jackets	☐	☐	☐	☐
sacrificial allowance	☐	☐	☐	☐
cathodic protection				
sacrificial anodes	☐	☐	☐	☐
impressed current	☐	☐	☐	☐
concrete lining	☐	☐	☐	☐
welded steel plates	☐	☐	☐	☐
ditto concrete backfill	☐	☐	☐	☐
protective row of piles	☐	☐	☐	☐
ditto concrete backfill	☐	☐	☐	☐

others ____ (describe)

The structure was repaired __________________________ years ago

The repair is still intact ☐

The 1st 2nd 3rd >3rd time of repair has failed after ____________________________ years.

Original extent of damage

- none ☐
- thinning ☐
- multiple small holes ("lace effect") ☐
- large holes ☐
- loss of backfill ☐
- collapse of quay surface ☐

Section 3 LOCATION OF ALWC (UK, EUROPE, WORLD)

Please mark the geographic position of ALWC on the map (*not included here*).

The results of this survey will help to identify suitable and economical solutions for repair and new design of steel marine structures and thus provide an important basis for the CIRIA Guide on The management of ALWC in steel marine structures, which will be published in Spring 2005. Thank you for completing this questionnaire. Please send it back before the end of April 2004 to:

Jim.Breakell@mottmac.com

Or print out a copy and send to: consider moving contact details to front

Dr Jim Breakell
Mott MacDonald Ltd
St Anne House
20-26 Wellesley Road
Croydon, CR9 2UL
United Kingdom

A6 PRELIMINARY ANALYSIS OF CIRIA QUESTIONNAIRE DATA

A. Estimated rates of corrosion

- averaged values over life of structure* to give minimum corrosion rates
- averaged values assuming longest induction period** to give indication of maximum corrosion rates.

Location	Age of structure	Assumed start of corrosion 1	Assumed start of corrosion 2	Assumed start of corrosion 3	Corrosion has first been observed in	Original thickness (mm)	Remaining thickness (mm)	Loss of section (mm)	Rate* minimum (mm/side/year)	Rate** maximum (mm/side/year)
1	1976	1981	1986	1990	1995	20.3	10.3	10	0.53	2.00
1	1955	1960	1965	1980	1995	22.5	0	22.5	0.56	1.50
1	1975	1980	1985	1990	1995	10.4	0	10.4	0.52	2.08
1	1975	1980	1985	1990	1995	10.4	0	10.4	0.52	2.08
1	1949	1954	1959	1964	1995	22.1	0	22.1	0.48	0.71
1	1966	1971	1976	1981	1995	22.1	0	22.1	0.76	1.58
1	1968	1973	1978	1983	1995	22.1	16.1	6	0.22	0.50
1	1981	1986	1990	1990	1995	22.1	18.1	4	0.29	0.80
1	1981	1986	1990	1990	1995	22.1	18.1	4	0.29	0.80
1	1978	1983	1988	1990	1995	22.1	10.1	12	0.71	2.40
1	1979	1984	1989	1990	1995	22.1	4.1	18	1.13	3.60
2	1995	1996	1996	1996	1998	21				
3	1992	1995	1995	1995	1999	19	12	7	1.00	1.75
3	1975	1980	1985	1990	1999	15.9	5.9	10	0.42	1.11
3	1990	1995	1995	1995	1999	19	12	7	0.78	1.75
3	1989	1994	1995	1995	1999	17.8	5.8	12	1.20	3.00
3	1955	1960	1965	1985	1999	15.9	5.9	10	0.23	0.71
3	1960	1965	1970	1985	1999	17.8	8.8	9	0.23	0.64
3	1970	1975	1980	1985	1999	13	2	11	0.38	0.79
4	1965	1970	1975	1985	2003	8.1	0	8.1	0.21	0.45
4	1965	1970	1975	1985	2003	8.6	0	8.6	0.23	0.48
5	1975	1980	1985	1990	2003	12.7	0	12.7	0.45	0.98
6	1995	1999	1999	1999	2000	9.8	5.8	4	0.80	4.00
6	1995	1999	1999	1999	2000	14	10	4	0.80	4.00
6	1995	1998	1998	1998	1999	14	13	1	0.25	1.00
6	1990	1999	1999	1999	2000	14	10	4	0.40	4.00
6	1990	1999	1999	1999	2000	14	13	1	0.10	1.00
7	1965	2001	2001	2001	2002	9	5	4	0.11	4.00
								Average corrosion rate [mm/side/year]	**0.50**	**1.77**

B. Corrosion observations

Feature	Number of Structures
No significant corrosion has occurred on the structure	16
Significant corrosion has occurred on the structure	27
The structure has been inspected for ALWC	38
ALWC corrosion has been identified	40
Corrosion has occurred, but unsure if it is ALWC	14
Corrosion is located:	
Just above mean low water springs	43
At mean low water springs	39
Just below mean low-water springs	35
Elsewhere	- some indication at well below LAT (6) - in dock 200 mm below normal minimum level - within a 1 m band above bed level which is +1.4 m LAT - within a 2 m wide band between 0.5m and 2.5 m above LAT. Worst attack seen around MLWN tide - full tidal range - bed level (4) - just below bed level where bed level higher than low water - just above mud line - at former bed level (-4 m) (2) - above bed - above bed some locations - down to bed.

C. Contributing Factors:

Based on the information supplied:

- the type of steel did not appear to have any influence as the incidence of ALWC occurrence was almost the same for both mild steel and high yield steel structures. However, almost 30 per cent of the respondees did not know the type of steel used in their structures
- some 78 per cent of the respondees noticed a random distribution of ALWC and the remaining 22 per cent claimed it occurred in groups (or in groups at regular intervals)
- ALWC was only reported as occurring in saline water with about 66 per cent of incidences recorded in seawater and the remaining 33 per cent in brackish water. No incidence was noted in freshwater. Approximately 50 per cent of the structures were located in ports and harbours (docks), the remainder were located in estuaries
- strong currents or a high level of water contamination may reduce the risk of ALWC occurrence
- approximately 60 per cent of structures were reported as showing signs of MIC
- orientation would appear not to be a significant factor influencing the incidence of ALWC.

D. Asset management:

A defined asset management system was in place in only 12 per cent of cases with about 69 per cent of responses where the structure is managed on an "as needs basis" (ie inspection prior to repair works). However, it would appear that records of previous inspection are being kept (which is the first step in stetting up an asset management system) with the majority of respondees having commenced records around 1995 (see graph below). A computerised way of managing inspection records (ie a database) was being used in the case of 33 per cent of the structures recorded, with the earliest entry dating from 1993.

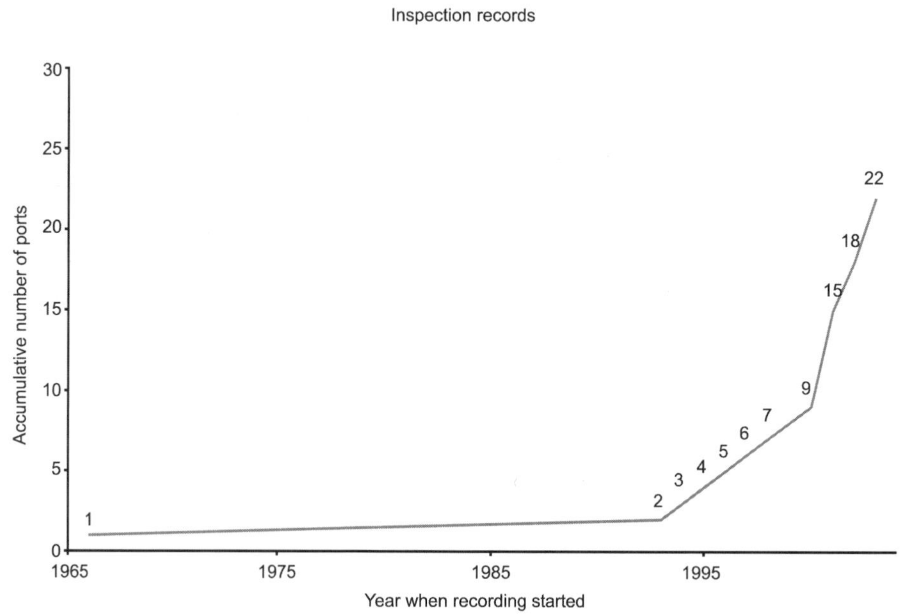

E. Inspection

Type of routine inspections and inspection intervals						
	every year	2 years	3 years	4 years	5 years	>5 years
Visual from surface	20	1	4	2		
Visual (boat)	14	10	22	9		3
Visual (divers)	15	8	2	4	2	3
Other testing	3	6				1
For combinations multiple fields were ticked						
Other types of inspection and measurements reported: - ultrasonic thickness testing - hammer testing for indication of residual thickness - diver inspection surveys - inspection in dry using limpet dams - monitoring and checking of impressed current CP system						
The structure is not inspected at regular intervals	5					

The type of reported routine inspections and inspection intervals are tabulated below:

F. Repairs and protection

- only 42 per cent of the ALWC damaged structures were repaired at the serviceability limit state ie when thinning of the section, but no holing, had occurred. However, the majority of repairs (58 per cent) was conducted when the structure was already holed, with 23 per cent of the structure in a very serious state of deterioration (loss of backfill or collapse of the quay surface), with a significant risk to health and safety of users prior to repair
- some 25 structures have either recently or previously been repaired, the repairs dating back over some 10 years. Some of the repair works have not been long lasting as eight per cent have already reportedly failed. In the case of one structure, it was necessary for it to have been repaired three times
- the majority of protection systems used comprised galvanic anode CP (38 per cent), coatings (30 per cent), and welded steel plates (15 per cent). The remainder comprised impressed current CP (8.5 per cent) and concrete encasement/jacketing (8.5 per cent).

A7 SURVEYING AND MONITORING

An ALWC inspection or survey can be carried out in a number of ways, and there are a variety of inspection and assessment techniques available to measure or monitor deterioration and obtain information on the mechanism of attack (Christie, 1999). The advantages and limitations of each are discussed in the following sections to allow the most appropriate method(s) and technique(s) to be used in a given application. Initial or periodic monitoring is regularly carried out using a boat survey with limited diving support to check condition in the submerged zone. Inspection and/or assessment techniques used are usually limited to visual and tactile with some thickness measurements and, in the case of monitoring inspections, electrochemical potential measurements. Similar access techniques are used for carrying out detailed surveys although the scope and extent of the survey will be significantly increased. Consideration should be given to the use of a limpet dam for this survey in cases where structural welded repairs are known to be required on the structure.

1.0 Means of access

The principal methods used for access to the ALWC problem for the purposes of inspection, repair or maintenance at and/or below LAT are:

- boat survey
- diving survey
- ROV (Remote operating vehicle)
- platform survey
- roped access
- limpet dam/dry survey
- steel sheet piled cofferdam.

However, some methods of access will not be suitable in some locations, for example due to restricted access, strong currents, low visibility, shipping movement, etc. The available methods should be considered for each application to determine which method or combination of methods will give the required results in an efficient manner.

Each method offers advantages, disadvantages and limitations which are briefly examined below, together with the practicality of application and an indication of the relative cost/quality of results.

In all cases, a reference system will need to be established before the survey starts.

a) Boat survey

ALWC damage on steel maritime structures is usually visible from a boat at low spring tides. A low freeboard boat is preferred to allow maximum access to the structure for close-up inspection.

Advantages	Disadvantages
• can be mobilised/demobilised at short notice.	• low tide dependant – short duration between ebb and flow tides. It is probable that only one or two tides every few months in any low spring period will occur during daylight hours, which are preferred for health and safety reasons. There are usually more opportunities in summer months with improved weather/sea conditions. Weather conditions are not usually a problem in harbours and small waves can be beneficial since they enable seeing below the static water level. Also, although daylight is preferred, it is possible to work at night with suitable lighting arrangements. However, weather and daylight are fundamental issues to be considered both in terms of health and safety and in the quality of site data obtained.
• minimum requirement permitting visual assessment and sample retrieval and thickness readings.	• shipping dependant.
• can survey large areas relatively quickly.	• holes sometimes not obvious due to deposits and marine growth.
• useful for producing still photographs and/or video survey.	• sample retrieval awkward or not possible.
• maximum accuracy.	• relatively cheap.

b) Diving survey

Factors that can influence diving operations include;

- water depth/temperature; although for most steel sheet piled wall structures, depth is not a major factor
- wave conditions
- current conditions; diving is possible providing currents are less than about 1.5 to 2 knots or by using a man basket
- safe access/egress
- visibility
- shipping and other port operations; divers should operate well clear of any shipping or marine craft movement
- obstructions (known and unknown), intakes/discharges.

A helmet mounted underwater camera connected via an umbilical cable to a surface monitor giving real time pictures and audio commentary should be considered to allow the survey supervisor to instruct a diver in a more thorough inspection and record the details accurately.

Advantages	Disadvantages
• can be mobilised/demobilised at short notice.	• working underwater can be awkward using bulky equipment in moving, deep or poor visibility water.
• not tide dependent.	• information second hand.
• can cover entire susceptible zone.	• special underwater tools required.
• can check piles right down to bed level.	• holes difficult to find by feel alone or not obvious due to surface deposits hiding the problem areas.
• underwater photographs and video possible but dependant on visibility.	• sample retrieval is difficult.
• detailed measurements and thickness readings possible underwater in conjunction with suitable reference system; 50-250 thickness measurements per day (depending on steel surface condition) with diving team (team of 4, with one diving at any one time).	• the quality of the survey is diver sensitive and, therefore, a competent diving team should be used with previous experience of ALWC and/or with suitable training.
	• can be expensive in relation to the quality of information obtained and highly dependant on the interaction between the inspector/diver and diving supervisor.
	• thickness measurements of pitted surfaces difficult, but usually obtainable, underwater; the minimum to expect is guidance on the adherence of surface deposits and whether excessively thinned areas are present determined by the use of a hammer or pick.

c) ROV (Remote operating vehicle)

These may be useful for underwater inspection on large structures and would minimise the safety risk.

d) Platform survey

Inspection from a platform could be provided in several ways including:

- a crane suspended man basket or cradle
- a floating pontoon or raft
- mechanical access platform lowered from the edge of the structure.

Which method to use will depend on site conditions and/or plant availability.

Advantages	Disadvantages
• can be mobilised quickly.	• low tide dependant.
• steady work platform when positioned.	• weather sensitive.
• sample retrieval and thickness readings possible as from a low freeboard boat.	• crane in constant attendance (if basket or cradle used).
• relatively inexpensive.	• only suitable for above waterline inspections and thickness measurements.
	• no access in dry to likely worst affected areas.

e) Roped access

Some specialist companies have experience in the provision of roped access (ie abseiling type) inspection and NDT of maritime structures (eg jetties, piers, marine vessels, etc) which may be of value for ALWC where other methods of access are inappropriate.

Advantages	Disadvantages
• can be mobilised quickly.	• low tide, weather and shipping dependant.
• visual assessment, sample retrieval and thickness readings possible as from a low freeboard boat.	• holes sometimes not obvious due to deposits and marine growth.
• can survey large areas relatively quickly.	• only suitable for above waterline inspections and thickness measurements.
• relatively inexpensive.	• sample retrieval awkward.

f) Limpet dam/dry survey

Inspection, repair and coating of a structure in the dry can be carried out simultaneously using a limpet dam supplied by specialist contractors. This is effectively an open sided box that is purpose built to mate against maritime structures of known profile (eg sheet piled walls, tubes, etc). It provides an effective access to both the tidal and submerged zones on a structure. Limpet dam widths are typically about 2-5 m, which can give access, for example, to a number of outpans on a sheet piled wall. Special limpets are made for corners and isolated pile access. A modified limpet dam (ie a circular submergible unit) has been developed that will enclose pier legs (New Civil Engineer, 2003). It is important to:

- ensure that the limpet dam is properly designed
- check the stability of any limpet dam on damaged walls
- confirm that the structure can bear the weight of the crane required to locate it.

Advantages	Disadvantages
• quick set up time; contractor's claim that access to the work site can be achieved within about 30 minutes from the time of lifting the dam into the water.	• setting up can sometimes be delayed due to problems, usually with wall characteristics. Complicated profiles can prove difficult since it may not be possible to adequately mate the limpet dam satisfactorily to the wall.
• surveys can be carried out quickly and efficiently.	• access down to bed level cannot be achieved; bottom 1-2 m inaccessible.
• access to depths up to 12 m below water can be achieved.	• relatively expensive due to the labour, equipment and planning requirements, particularly for complicated profiles. Costs can be reduced if the client can provide some of the necessary on-site resources (eg crane, plant or labour).
• permits free access to and allows direct inspection of the critical area under dry conditions, irrespective of tides. Therefore, the quality of data is superior to that obtained by diving inspection.	• in danger of being dislodged if struck by shipping. Therefore, need exclusion zone around operations.
• holes, leaks, defects and corrosion are visible and measurable.	• only 2-5 m per limpet dam can be inspected at any one time.
• allows the client to attend each set-up in order to fully appreciate the condition of the structure.	• mobilisation time can be long. Reliant on small number of firms having the requisite experience.
• reveals far more about the nature and behaviour of the structure in question than by any other means – particularly useful if limpet dam based repairs are anticipated in the same area.	• there can be problems keeping the dam dry if a sheet piled wall is badly holed and there is porous material behind. There is increasing difficulty and expense in carrying out such works as the size of holes increases.
• blast cleaning of the wall surface in the submerged zone can be achieved prior to coating application.	
• allows collection of debris (eg dust, grit, rust, etc) and paint particles during surface preparation and coating application.	

g) Steel sheet piled cofferdam

A steel sheet piled cofferdam can be formed by driving an enclosure of steel piles around the structure or element and dewatering the area to allow dry access. This is less flexible and somewhat more expensive than installing a limpet dam, and is unlikely to prove viable unless the cofferdam is already there for another purpose and any maintenance or repair resulting from the inspection can be carried out at the same time.

2.0 Inspection and assessment techniques

The inspection and assessment techniques applicable to the identification and measurement of ALWC in the low water and/or submerged zones are considered below.

The most appropriate technique(s) for a specific structure will depend on the particular circumstances and requirements.

a) Visual and tactile:

The following types of visual inspection should be specified:

- general visual inspection; quick, inexpensive form of inspection allowing identification of "orange patches" and holes at some distance from the structure. The outer marine fouling layer is not removed which may limit the detection of ALWC. Holes just below low water can often be detected by the surface movement of water caused by the escape of water behind the wall
- close visual inspection; involves visual and tactile examination at and around the previously identified ALWC patches. This form of inspection usually involves removal of the surface deposit using mechanical means. At ALWC patches the "orange" surface deposit can easily be removed by wiping by hand, allowing residual thickness and/or pit depth measurements to be taken on the substrate steel. In contrast, the surface deposit away from the ALWC patches is likely to better adhere to the steel surface, although a hand scraper should remove it as required. However, removal of such well adhered surface deposits should be kept to a minimum (see Chapter 5, Section 5.2.5)
- tactile inspection; this relates to diving inspections where there is little or nil visibility and the diver undertaking the survey relies only upon sense of touch and perhaps a hammer to investigate the structure. Tactile inspection can only be relied upon to locate holes greater than 25-30 mm wide. Under such poor visibility conditions, local areas of the structure can be exposed for visual inspection by using a clean water box to improve visibility. The ALWC inspection, with the need to touch and remove surface deposits, can itself have visibility problems.

b) Visual recording

The following techniques provide a permanent visual record of findings.

- still photography; this should include a reference number, date, a scale rule for close up photographs and colour chart; the use of colour photographs in the report will greatly improve the visual appreciation of ALWC. Numbered photographs showing general views of long sections taken at distance from the structure are also useful for illustration purposes
- close circuit television (CCTV) cameras; relays information to a monitor or recorder on the surface. Videos should have the location and orientation clearly described and any other descriptive text included, as required. Such equipment can be handled underwater by divers or attached to a pole above water allowing an inspector to lower the equipment down a pile and visually record the surface.

c) Measurement equipment

- *tape measures, rulers and micrometers:* Micrometers are often used to measure residual thickness on free edges such as the flanges of H piles
- *hammer or pick:* A hammer would be expected to hole through steel at areas where the residual wall thickness is 1 mm or less and dent at 2-3 mm or less. Deliberate perforation is not recommended as holes can complicate the repair process later on. An experienced inspector is often able to estimate remaining thickness by the sound of a hammer impact of given weight
- *pit depth gauges:* For measurement of the depths of individual corrosion pits relative to the surrounding surface away from the ALWC; pit depth instruments (eg

mechanical depth gauges, needle gauges, etc) are commercially available

- *illuminated magnification:* Low magnification (x10, x25 or x50) optical microscope with built in graticule to provide an estimate of the width of pits and, with suitable magnification, their depth and configuration and whether saucer or "omega" shaped. This is important before making recommendations for surface preparation prior to coating application and also to identify the type of pitting across the surface
- *callipers:* Used to provide an accurate measurement of thickness; operate through a small (minimum 12 mm diameter) hole drilled through the steel surface which will subsequently require repair
- *profile gauges:* Needle instrument which when forced against the surface will deform to match its profile
- *impression mouldings:* Used to obtain a replica of surface profile
- *ultrasonic thickness equipment:* Used to establish residual thickness measurements on steel structures. Two types of ultrasonic thickness (UT) meter are available
    1. Digital thickness meter with LCD readout; this is the instrument which is most commonly used since readings are easy to take, the readings can be stored within the meter itself for access later and equipment is available which allows measurements to be taken underwater.
    2. Flaw detector; this instrument uses an oscilloscope to display the measurements and can easily detect any through thickness defects within the steel wall and/or localised metal wastage from the back surface.
- *pull off adhesion tester:* Destructive method to provide adhesion of coating to substrate.

3.0 Ultrasonic thickness surveys

Areas to be measured (mainly at "orange patches" but with a limited number of readings on surrounding piles) should first be cleaned locally of all surface deposits before taking the readings. The thinnest areas can usually be seen when the surface deposits have been cleaned off. A coupling agent will be required for above water measurements to ensure good contact of the probe with the steel surface. The seawater itself typically acts as coupling agent for underwater measurements. In high turbidity waters, a sufficiently long connecting lead from the probe allows readings be taken on the meter above water. On-site calibration checks using shims of known thickness should be carried out on the meter before and after each set of readings.

Residual thickness measurements during detailed surveys should typically be obtained using a grid system (ie both along the structure and from the top to the bottom of the structure) across the full surface of the steel elements being investigated down to bed level. Clearly, it is important that that there should be a concentration of measurements at and around the LAT zone, which is the area of greatest risk for ALWC. It would also be useful to check the thickness in the most highly stressed parts of the structure. Details on the number of vertical and horizontal measurements to be taken can only be for guidance purposes, as the actual number required will vary from structure to structure. In general, vertical measurements should be obtained at 200 mm to 300 mm intervals within a band measuring approximately 1200 mm either side of the low water level. Horizontal measurements across or around the structure should take into account the "classic" characteristic patterns of ALWC damage. A typical frequency for horizontal measurements on different pile geometries in the critical zone is as follows:

- for steel sheet piling at least eighteen measurements per vertical interval per pile should be obtained, six in the outpan flange, six in the inpan flange and six in the web

- for tubular steel pipe piles a minimum of three measurements at four locations, equally spaced (ie at 90 degree intervals) around the pile circumference should be obtained. The location axis should be orientated so that one location is facing toward the seaward direction
- for steel H piles at least twenty measurements per vertical interval should be obtained, eight on each flange and four in the web.

It can be difficult to take readings on a pitted or previously coated surface. The thickness value reported to the nearest millimetre for the worst areas of general thinning should be derived from a single minimum thickness measurement taken within the ALWC patch after removal of surface deposit rather than an average of a given number of measurements within this location. In addition a value for the deepest pit should also be obtained within the ALWC patch using a pit depth gauge. Alternatively, it is possible to grind flush (using hydraulically or pneumatically operated tools) any badly pitted areas before ultrasonic thickness measurements are taken to obtain this minimum reading but, in practice, there is usually not enough time to do this. In contrast, the thickness value reported as representative of the remaining sound metal should be an average number of ultrasonic measurements reflecting the more uniform and reduced corrosion occurring away from these local areas of thinning. Where thickness readings are attempted but are considered unreliable this can be denoted on the survey drawing by "N/R" (no reading). Where readings cannot be obtained, for example due to restricted access, then this can be denoted on the survey drawing as "N/A" (not applicable).

4.0 Electrochemical surveys

Measuring electrochemical potentials on steel maritime structures can be used to tell whether a macro-corrosion cell or possibly stray current is operating on the structure. Measurements of structure to seawater potential can be taken:

- from the surface by (1) a portable reference electrode and voltmeter or (2) a permanently installed reference electrode and voltmeter
- using divers with hand held potential monitoring devices.

Reference electrodes should be Ag/AgCl/seawater or Zn/seawater type and suitably calibrated prior to use.

The two techniques differ mainly in:

- *the way that electrical contact is made to the steel surface*: In the case of the remote system, connection is made to the structure above the water line. With the diver held system it is made through point contact by the probe itself (Balmer, 1984). Obviously monitoring from the surface is a much lower cost option
- *the accuracy of the measurement made*: The remote system provides an average potential value from over a large area of the structure, while the diver held system measures potential at a precise location.

As with the ultrasonic survey, readings are usually taken on a grid pattern across the whole surface of the structure while it is immersed. The results obtained can then be compared with those obtained from a residual thickness measurement survey to determine any correlation between the variations in electrochemical potential distribution and metal wastage pattern seen across the structure (Vrelust, 1992).

Stray current monitoring can be undertaken by either interrupting impressed current cathodic protection system or monitoring during shutdown periods if rail or welding interference is suspected. Bimetallic factors are more difficult to locate with this technique unless it is possible to isolate structures from each other and then install temporary bonds for testing and monitoring.

It is important that divers working in the vicinity of impressed current CP systems follow national and local guidelines for the safe use of electricity underwater.

5.0 Ongoing monitoring

In order to monitor the rate of ongoing metal loss due to ALWC, further thickness measurements need to be taken at regular intervals. It is important that these ongoing measurements are made at the same location. However, it should be recognised that the results obtained may be influenced by the frequency of measurements because of the requirement to remove surface deposits prior to measurement.

It should also be recognised that ultrasonic thickness surveys can often show considerable scatter in individual measurements for a variety of reasons, particularly when they are taken on uneven and deeply pitted surfaces such as those often found in the low water zone. Also some variations should be expected when comparing readings taken by different inspectors (since measurements are often influenced by judgment on the part of the inspector) particularly when measurements are taken at different times or when different ultrasonic equipment is used. To minimise such inconsistencies and subsequent difficulties in interpretation, it is important that a consistent method of approach be used as far as possible in the measurement method and in the recording of information between monitoring intervals (see Section 3 above and Chapter 2 Section 2.4).

6.0 On-site sampling/analysis

While major areas of corrosion and deterioration are likely to be evident from visual inspection, sampling followed by on-site testing and/or sampling and storing for subsequent more detailed laboratory analysis of liquids and/or surface deposits are necessary to confirm microbiological activity (ie MIC) and the risk of ALWC occurring at any particular location.

A certain degree of biochemical knowledge is needed to effectively perform on-site sampling and testing. While this need not rule out the engineer or surveyor carrying out this work, appropriate skills, training, methods and materials should be acquired and used if it is to be done effectively. If sample retrieval is to be completed then this should be identified when planning the survey as a steady work position will be needed (see Chapter 5, Section 5.2.4).

Several types of testing that can be conducted on site to provide useful information on:

a) *The presence of SRB*; the most common method is to culture the bacteria in special nutrient broths which are specific to the kind of bacteria tested. In the case of SRB, testing using bottles containing marine API-RP-38 media is recommended although other commercial test kits are available, usually based on enzyme tests. A useful guide for engineers required to carry out such tests is in *"Review of Current Practice for Monitoring Bacterial Growth in Oilfield Systems"* Corrosion Control Engineering Joint Venture Task Group E1/1, Document No.001/87. For this type of testing a liquid sample is taken using a sterile syringe from behind or within the black sludge at ALWC patches. Comparative tests can also be conducted on samples from the bulk seawater and/or liquid from the mud zone. Rapid enumeration kits are commercially available for a variety of organisms such as SRB to be made on site. As a general rule, the more rapid the technique in producing a result, the lower its sensitivity will be.

b) *The nature of the black surface deposit* ie confirming that the black deposit is iron sulfide rather than iron oxide (ie Magnetite). Iron sulfide in the corrosion product is often a sign of SRB activity. A simple test involves acidifying the black deposit with dilute hydrochloric acid. Effervescence with the liberation of hydrogen sulfide (smells of rotten eggs) confirms that the black deposit contains iron sulfide. Lead acetate paper should be used as an indicator for hydrogen sulfide. Caution should be taken when performing this test not to inhale hydrogen sulfide as it is highly toxic.

c) *Laboratory testing*; certain forms of iron sulfide are reported to be pointers to SRB activity (ie the presence of Mackinawite, Smythite and Greigite). In order to determine the chemical nature of the previously identified iron sulfide, samples of the black deposit will need to be removed in a specialised manner under oxygen free nitrogen. Once removed, samples should be stored in airtight containers clearly marked with their reference point of origin pending analysis under laboratory conditions. Such specialist removal and storage procedures may be considered beyond the scope of the normal inspection. It should be stressed that any exposure of the sample to air (during the sampling, storage and analytical stage) will cause rapid changes to its chemical composition (oxidising the iron sulfide to iron oxide and elemental sulfur) rendering the analysis effectively useless.

A8 EXAMPLE OF TYPICAL DETAILED CONDITION APPRAISAL

- introduction (including a summary of the clients brief and of the inspection undertaken)
- limitations and exclusions (the 'exclusions' are not necessarily part of the factual report, but can be part of the asset management strategy)
- description of structure
- definition of numbering of structural elements
- risk assessment
- previous reports
- inspection procedure
  - personnel including qualifications and experience
  - visual
  - photography
  - thickness measurements; types of probes used
  - electrochemical potential measurements; type of equipment used
  - any other inspection techniques used.

- findings including
  - completed inspection data sheets
  - plans or drawings (annotated with inspection notes from visual survey) relating to area numbers or a grid reference system
  - dimensions, still photographs, video film, etc
  - results of any destructive or non-destructive tests cross referenced to location and level and date of such tests
  - samples taken.

- structural assessment
  - corrosion rates/assumptions
  - analysis of the information or data showing what relationship exists to the original design or information obtained from previous inspections.
  - re-analysis of structure
  - reassessment of durability and strength.

- recommendations for
  - prediction of remaining serviceable life
  - the need for any further investigations
  - remedial actions; options for low cost, short term solutions or higher cost, long term durability; immediacy of repairs; in the case of coatings, details of specification including type and preparation standard
  - cost estimates.

A9 A MODEL GENERIC RISK ASSESSMENT (Christie, 1999)

Note that items can either be extracted for suitable action or be added to for the development of a dedicated risk assessment, as appropriate.

Assessment date	**Initials**	**Number of pages**	**Review no**	**Date**	**Initials**
Description: **STEEL PILE CORROSION SURVEY – GENERIC**					

Hazard	**Risk** H – High, M – Medium, L – Low	**Controls**	**Notes**
Immersion	Drowning	Buoyancy aid	Water usually very cold or near freezing, effect rescue/recovery procedure immediately
	Hypothermia	Change of clothes, blanket	
	Swept away	Life-buoy, safety line	
Weather	High winds		Use weather forecast Rapid changes can occur Establish stand-down criteria
	Sun		
	Rain	Avoid over-exposure, dehydration	
	Snow, ice	Avoid electrical storms	
	Fog	Avoid numbness and chill – thermal PPE Abort	
	Darkness	Back-up lighting	

Cofferdam	Confined space	Define safe system of work	Ref. Confined Space Regulations
		Emergency procedures	
	Design	Fitness for purpose by qualified source	
		Competent person	
	Installation/removal/alterations		Competence
		Diving at Work Regs. 1997	Qualifications
Diving	Variable risks (subjective)		Health certification
			Equipment checks and certification
			Diving log/records maintained
			Risk assessment
			Safe working system
			Emergency procedures/first aid
			Communications
			Project plan
Water movement	Flowing water	Life-line	
	Wave/chop	Adequate, stable craft	
	Shipping wake	Radio contact with control	Divers – display flag
		Lookout	
Shipping	Moving	Hi-visibility	
		Vigilance	When on quayside
		Vigilance	Waterborne craft, crane ops
Quayside operations	Moorings – trip hazard		
	Moorings – obstacle	Hi-visibility	Establish prior if site expected to be busy. If so reschedule for quiet period
	Cargo handling, vehicles	Keep safe distance	
Isolated location	Isolation – emergency	Stand-bye person, radio/phone	Dedicated risk assessment in extremes

Private areas *Structural examination* *Outfall pipes* *Obstacles*	Designated Hazardous General Head injury Eye injury Hand injury Over-reaching Deluge Loose fenders, fittings Underwater obstructions	Work permit, site specific procedures Hard hat, bump cap, lanyarn for tools Safety glasses, goggles H/D rubber gloves Keep body within confines of craft/platform Isolate Lock off Avoid Avoid Diving risk assessment	Obtain and respect procedures Desk study prior to establish status and risk Note and report for repair
Lifting operations	Variable risks (subjective)	Lifting Operations and Lifting Equipment Regs. 1998	Certification of lifting appliances in place All lifting appliances marked with Safe Working Load Trained and competent operator and slingers Risk assessment Avoid use of man cradles if possible

A10 CORROSION PROTECTION METHODS

1.0 COATINGS

External protective coatings for maritime applications may be organic, inorganic or metallic materials that may be applied by spraying, brushing, dipping, wrapping or co-extrusion

1.1 Suitability of coatings

The type of coating used will depend on various factors including:

- *economics*; the costs per square metre of applied coating will vary depending on the coating system used. In the case of organic coatings the costs per square metre are probably not too dissimilar for the types of organic coating systems likely to be used on a sheet piled structures
- *environmental considerations;* in the last decade, the use of solvents has been significantly reduced in response to the Environmental Protection Act (1990) and the subsequent issue of PG6/23 (CIRIA, 1997). VOC compliant coating systems that have VOC contents meeting the current requirements of PG6/23 should, therefore, be specified (for factory applied coatings). A coating should not be specified which contains bitumen or coal tar because the polycyclic aromatic compounds in tar and tar derivatives are known carcinogens (ie cancerous)
- *surface condition of the steel surface to be protected*; the performance of any coating will be determined by the quality of the surface preparation (see Section 1.4 below) although some coating systems are said to be more surface tolerant
- *ease of application*; for example, the high build solvent free organic coating systems require suitably trained applicators using specialist application equipment compared with the easier application of solvent based systems
- *resistance to impact and/or abrasion*
- *degree of performance required in a particular environment*
- *time to first maintenance*
- *ease of repair*
- *curing (or drying) time;* for coating application within the splash and/or tidal zones, the time interval between low and high tide is relatively short and the coating will be quickly exposed to the seawater unless applied under dry conditions. The coating should ideally be able to be applied over a wide range of humidities and temperatures
- *compatibility*; with supplementary corrosion protection methods. For example, the coating will need to have been tested by the manufacturer for acceptable resistance to cathodic disbondment where CP is (or likely in the future) to be used. Written confirmation from both the CP system designer and the coating manufacturer should be obtained to demonstrate compatibility. Cathodic disbondment tests to BS3900-F10, or equal standard, can provide corroborative data
- *health and safety*.

I.2 Application

Coatings can be considered at various stages in the lifetime of a steel piled structure (Van Damme, 1999):

1. **Before driving the piles**
   - coating applied under factory conditions; quality control is far better than on-site work
   - likely transport and handling damage
   - vibration caused by driving the piles reduces adhesion of the coating to the steel substrate, particularly near the pile head
   - damage to coating can occur at the interlocks (ie clutches)
   - mechanical damage to the coating can occur as a result of the abrasive action of the soil material in the mud zone (for this reason, this region is normally left uncoated). Although ALWC would not be expected to occur below the mudline, MIC is still possible if the muds are heavily contaminated
   - require limpet dam to repair coating damage on surfaces in submerged zone; the cost will be similar to coating after driving the piles (see (3) below). If CP is also being applied coating repairs in the submerged zone may not be essential although disbonded coatings can produce current shielding effects.

2. **At the time of the construction**
   - an approach where "the factory" is brought to the site and a production line for grit blasting and coating the piles arranged
   - provides the required quality control
   - reduces the risk of pre-driving mechanical damage to the coating
   - other problems as for (1).

3. **After driving of the piles**
   - coating applied in dry using limpet dam
   - sheet piles have a smooth surface easy to coat with a minimum of preparation.

4. **Maintenance works a number of years after construction**
   - maintaining coated structures is more difficult to achieve successfully than coating of new steelwork
   - increased difficulty in achieving required surface preparation
   - the required quality control is difficult to achieve
   - increased cost
   - coatings need to be tolerant to application to poor surfaces under poor conditions and be compatible with CP if used in combination
   - may be necessary to remove components (eg fenders) to gain access to surfaces.

5. **Repair works followed by maintenance**
   - can produce accelerated localised attack at "holidays" or damaged areas of the coating if only applied to the repair works.

6. **Steel sheet piled structures that have to be rebuilt**

1.3 Types of coating

The generic types of coating systems used for protecting steel in the marine environment and their track record and likely suitability in protecting specifically against MIC are discussed below.

a) Organic coatings

High performance organic coatings, correctly applied to a suitably prepared substrate (ie to the required surface cleanliness, anchor profile and free of visible and invisible surface contaminants in accordance with the coating manufacturer's recommendations) have a long and successful track record for the protection of steel structures, continuously or partially immersed in sea water. Increases in effective life of typically between 8-30 years can be achieved using this method depending on the quality of surface preparation achieved and the coating system used. For maximum effect, quality control and inspection both during and after application by an experienced and competent coating inspector (qualified to at least Institute of Corrosion Coatings Inspector Level 2, but preferably Level 3), representing the owner, is recommended. Regular inspection and maintenance regime needs to be adopted particularly for those areas susceptible to ALWC to maximise the effective life of the coating.

It is essential that coatings used in such a harsh environment have good adhesion, limited moisture permeability and high durability. There are many commercial coating systems available that meet all the necessary requirements for marine exposure and, as such, they are not new or special products. For the present service, the coatings should also have satisfactory resistance to biodegradation and chemical attack by microbial activity (eg hydrogen sulfide, organic acids, etc). Coatings should be formulated with resins, pigments, plasticisers, etc which in themselves cannot be used by microbiological organisms as a food supply. Some antifoulant type coating products containing biocides have been developed specifically for protection against MIC although their effectiveness is relatively short term. Certain existing coating systems have been recommended for the protection against ALWC (Marsh, 1998).

Organic coatings are susceptible to mechanical damage (eg from impact and abrasion, bow thrusters and propeller wakes, dredging, etc) although some more abrasion resistant formulations are commercially available, for example the high film build, two pack, cross linked coating systems (particularly those pigmented with flakes) or three pack coatings similar to those used for piling protection on the Thames Barrier. Damaged areas in coatings may become anodic to the remaining surface and intensified corrosion may develop at these areas unless they are repaired regularly or used in combination with a well designed and maintained CP system. Significantly disbonded or blistered coatings can, however, be problematic since they can produce a shielding effect, preventing the protective current from reaching the substrate steel. It is also necessary to consider the implications of partial coating on steel surfaces which could result in increased corrosion rates on other parts of the structure.

The principal generic types of coating used include:

Coating system	Comments	Track record for protecting against MIC
Solvent free (ie 100% solids), high build epoxy ***Also solvent less (eg 90% solids)***	Usually a two pack, solvent free, high build, single coat (400 to 600 µm DFT) epoxy system which is applied under dry conditions on grit blasted surfaces at least to Sa 2½. Specialist equipment is required for application of hot applied solvent free coatings. As above, but contains some solvent which allows easier application and film formation. A polyamide epoxy is general purpose and is designed to be readily recoatable, whereas the amine adduct epoxy type is harder, more chemically resistant but significantly less recoatable after it has cured (eg in one to two weeks).	Solvent free high build epoxy coatings have been used to protect steel pilings (eg Sovang, 1998; Vrelust, 1992; Van Damme, 1999). Also see case study 1 (Appendix 6.2) for practical example. Amine adduct epoxies will likely perform better than the polyamide types in severe ALWC conditions since the latter systems are more susceptible to biological attack.
High build, glass flake epoxy	A single 500 µm DFT coat of glass flake epoxy applied under dry conditions onto suitably prepared surfaces can be a very durable coating for the protection of steel sheet piling against ALWC. Important factors in glass flake coatings are the type of pigment (lamellar or spherical), size distribution and the pigment content. Glass flake epoxy materials are prone to rapid chalking and fading on exposure to UV light. If appearance is important, then they are usually overcoated with a polyurethane (PU) finish. Most glass flake epoxies from the major manufacturers have recoatability limitations and would, therefore, appear to be amine adduct epoxies.	Gave the best performance overall in recent tests (Marsh, 1998; Moulin, 1999) comparing the performance of 25 different marine paint formulations. The choice of coatings used in the test programme was made taking into account (1) all the major generic types of coatings, (2) the practice of various countries utilising steel piling in Europe and (3) the appropriate environmental legislation. Not all glass flake coatings will necessarily give the required long life performance. The correct amount of appropriate particle size and distribution of glass flake pigment within the coating is important.
High build glass flake polyesters	Typically a two-pack rapid cure, polyester glass flake coating although three pack systems are also available. Applied as a single coat (350 to 600 µm DFT). Surface preparation is a contamination free, blast cleaned surface to Sa 2½ near 3 or equivalent. Can be airless sprayed or roller brush applied. Some available that will cure at temperatures below freezing with full cure obtained within four days at 20°C.	Thick glass flake polyester coatings have been used to protect steel pilings (Kumar, 2002). Glass flake polyesters cure more quickly and are more light stable than glass flake epoxies.

Vinyl co-polymer systems	Single pack, rapid cure and surface tolerant coating system. Typically applied as a multi-coat system (225 to 250 μm DFT)	Has been used in maintenance role on steel structures in tidal applications (eg on jetties) and on immersed surfaces (eg ship's bottoms).
Tar free, moisture curing polyurethane (MC PU).	A multi-coat system comprising a MC PU primer (with reduced zinc dust) and intermediate/topcoats with a minimum DFT of 450-500 μm. Several coats can probably be applied during one low tide. Use a primer without zinc dust additive if used in combination with CP, to maintain a high insulating property.	Not widely used under immersed seawater. Has been used in sheet piling application (Schwindt, 1999).
Elastomeric	A chloroprene sheet lining having a thickness of 3-4 mm bonded to the steel surface. This is an expensive but potentially very durable coating if applied correctly.	Main use has been in the oil and gas industry as a high temperature coating for protecting submarine pipelines and as enhanced coating protection on risers as they pass through the splash zone. Has been applied to tubular piles and steel sheet piles in one UK harbour. Possible problems repairing the lining under wet conditions.
Three layer coating system	Applied by co-extrusion process to tubular piles under factory conditions typically involving the application of (1) a fusion bonded epoxy (FBE) anti-corrosion coating, (2) bonding agent and (3) a 3-4 mm layer of HDPE. This can be a very durable, high performance system. CP neither suitable nor necessary. Repairs to damage occurring before installation are relatively easy and cheap. Repairs to damage after installation can be dealt with by wrapping techniques.	Widely used in the oil and gas industry for protection of submarine pipelines. No known track record for use as a coating for tubular piles in marine applications.

b) Inorganic coatings

Coating system	Comments	Track record for protecting against MIC?
Cementitious	A single high build (eg 2000 μm DFT), surface tolerant, rapid curing, epoxy modified, cementitious coating. Can be installed between tides to damp surfaces and will cure under water. As for concrete facings, they can chemically passivate the substrate steel due to their high alkalinity.	Limited track record to date of use on steel piled structures (Anon 2003).

c) Metallic coatings

Coating system	Comments	Track record for protecting against MIC?
Metal spray	Metal spray the whole surface of the steel piles before installation with aluminium, overcoated with suitable sealer used to penetrate and seal the porosity. Partial sealed metal spray coating can result in rapid loss of the aluminium coating due to bi-metallic action with any adjacent uncoated areas in electrical contact. Metal spray coatings can contain high porosity which is not always possible to seal and, in such circumstances, hidden premature corrosion of the base metal can occur. A compatible topcoat paint system can be applied over the sealed metal spray to provide additional protection to this coating. Only use on large flat areas. Not suitable on areas which are difficult for the applicator to achieve the specified minimum thickness without multi passes.	Metal coatings have been used successfully on immersed structures such as jacket legs of production platforms in the oil and gas industry. The latter were large diameter tubular sections which were automatically coated and free from porosity in the bulk of the coating. Also, limited use on steel sheet piles (AISI, 1981; Kumar, 2002; Hou, 2003). Slow and costly process.
	Possible problems if used with impressed current CP since they may suffer accelerated attack from the increased concentration of hydroxyl ions generated particularly in those areas where the current density is above the required level. Should be OK with galvanic CP.	
"Duplex" coatings	Hot dip galvanised surface overcoated with an organic coating. The preparation of hot dip galvanising and the pre treatment of these surfaces prior to coating should be of the non water-soluble type and the coating applied prior to the formation of any zinc salts. The use of direct application primers may be considered but these coatings should have a proven track record in immersed conditions.	Not widely used in marine environments and only on new build structures.

d) Others

Coating system	Comments	Track record for protecting against MIC?
Wrappings/tape protection systems	Typically petrolatum gel tape wrap systems. They are surface tolerant systems and can be installed both above and below water. Installation tends to be labour intensive, usually requiring up to four operations; (1) initial cleaning of surface, (2) application of paste, (3) taping up and (4) fixing the outer protective wraparound jacket with a strapping, fastener or bolted system. Surface pit depths greater than around 2mm normally require the use of an initial primer or filler. The wraparound jacket should be applied in close contact with the tape to prevent water ingress and movement beneath the jacket with tidal movement.	Well established for application to tubular piles particularly in the splash zone. Not appropriate for angle sections. The petrolatum tape system is reported by the manufacturer (Smith, 2002) to incorporate both an inhibitor and biocide to minimise the risk of corrosion and microbiological attack. Also see case study 4 (Appendix 11) for practical example.
Anti-foulant coatings	These coatings contain biocide (most commonly oxides or other compounds of copper or tin) which is slowly released from the coating over time to inhibit growth of micro-organisms. Conventional anti-fouling coatings probably have a life expectancy of approximately one year. They do not, therefore, offer a long-term solution to the problems associated with ALWC.	Previous track record of use in marine environment but have environmental implications leaching potentially toxic materials into the environment.
Underwater cured epoxy coating	Typically high build, solvent free epoxies that cure underwater. Such materials typically have consistency of a thick grease or putty and are usually applied (sometimes with difficulty) by divers using a mitt or by trowel.	Primarily developed for maintenance purposes, typically as a low cost protection method on old, heavily pitted steel pilings (AISI, 1981) and can be applied using divers on small or large areas (Windsor, 2000) or in the dry, using a limpet dam, allowing it to continue curing underwater. May, therefore, have some application as low cost option for local protection at ALWC patches following easy removal of surface deposit revealing shiny steel substrate beneath.

1.4 Surface preparation

On-site application of high performance coatings is never easy and will always be inferior to factory applied coating systems. In order to ensure long term performance of on-site applied systems, there should be no compromise on the quality of surface preparation and application. Correct surface preparation is necessary for good adhesion and effective corrosion prevention. The surface preparation should be a minimum of Sa 2½ (ISO 8501-1/SSPC-SP 10/NACE No 2) and the application of the coating system should be effectively carried out. Surface preparation (including removal of surface salt contamination, corrosion product salts within pits, residual hydrocarbons, etc) should be performed under dry conditions using limpet dams. Ordinary dry blasting will in many cases be inadequate, although to the untrained eye may appear to be satisfactory. Spent grit, removed surface deposits, etc should be collected to prevent contamination of the water or mud zone.

The following on-site surface preparation procedure under dry conditions is suggested for existing structures with MIC contaminated surfaces:

- initial high pressure (HP) or ultra high pressure (UHP) water jetting using fresh water to remove deposits and residual salts (ie chlorides, sulfates, etc) from surfaces and within pits. UHP can also remove loose mill scale. The surface should then be washed with a 1 per cent solution of industrial hypochlorite (ie a biocide) with due regard to health and safety and environmental requirements. The use of steam cleaning or solvent cleaning to remove organic contamination (ie oils, greases, etc) should also be considered, where necessary
- the area should then be thoroughly rinsed with fresh water and allowed to dry
- blast clean the surface to a minimum of Sa 2½ and create the required anchor profile for the coating system to be applied (up to 120 μm Rz - the average peak-to-valley height of a rough surface)
- fill heavily pitted areas with weld metal or epoxy filler (stripe-coat pits with chosen coating before applying filler) that is compatible with the coating system to be applied
- apply coating system in accordance with manufacturer's recommendations.

Residual salts on surfaces and within pits, even at low concentrations, can stimulate osmotic blistering and corrosion beneath a coating which can ultimately cause rupture of the film. For a high performance coating, the concentration of residual surface salts should be specified at less than 1 microgram per square centimetre in the tidal and immersed zones although slightly higher levels can be permitted above the high water mark. It is important to note that in a marine environment, salts are continually being re-deposited onto steel surfaces (or between coats for multi-coat systems), so even after washing/rinsing of surfaces with fresh water, chloride levels can be relatively high particularly (1) if measures have not been taken to protect the steel surface, (2) there is a significant delay in applying the coatings or (3) the chloride level in the wash/rinse water is high. Salt deposits would not be visible on a surface until they are at a concentration above 750-1000 micrograms per square centimetre. Methods are available to identify the level of contaminants present on a steel surface, either (1) for specific salts (eg chlorides, sulfates, etc) although these can sometimes be time consuming and difficult to implement on site and (2) those for measuring general salt levels on unpitted surfaces, such as with a conductivity measuring gauge.

Surface preparation can also be achieved underwater using water jetting, abrasive blasting or mechanically powered tools with due regard to environmental considerations. Where blast cleaning to Sa 2½ is not possible, consideration should be given to more "surface tolerant" coatings which may be applied over less well prepared surfaces using less rigorous cleaning methods (eg hand tool, power tool cleaning, etc) although the resultant performance will be inferior to a high performance coating suitably applied. It is important to be aware of exactly how "surface tolerant" specific coatings may be in practice. Even though the high pressure water jetting process (ie in the absence of abrasive) does not create a profile on the steel, it is capable of exposing previously blast cleaned steel surfaces under old coatings sometimes producing an adequate surface for the application of moisture curing tolerant coatings.

2.0 Cathodic protection

Cathodic protection is an electrochemical method of corrosion control. Its use is mandated in the offshore oil and gas industry by Statutory Instrument and Classification Societies for the protection of immersed surfaces of fixed and mobile marine structures either alone or in combination with coatings. The emphasis in this section is on the practical aspects of CP. If properly designed, it should be an effective way of protecting the entire submerged structure from any form of corrosion up to approximately the mid tidal zone. It will not protect steel surfaces in the splash or atmospheric zones. It has been estimated that CP is probably the most economic form of protection if the residual, life of the structure exceeds 20 years (Wyatt, 1999). CP has been used for protecting maritime steel structures in the UK such as jetties since the early 1950's and for steel sheet piling in other countries such as Scandinavia, Germany (Reinhard, 1998), Japan (Kazuaki, 1985) and the USA (Kennedy, 1967). Cathodic protection can be installed or planned for when the structure is built or retrofitted to an existing structure (Marsh, 1998).

It is important that a CP system is carefully designed in order to be capable of delivering and distributing sufficient DC current to maintain the submerged sections of the structure in a non-corroding condition for a specified period of time. The are two types of CP systems:

- galvanic anode CP; ie coupling the structure to a more electronegative metal; typically uses zinc (Zn) or zinc indium activated aluminium alloy (Al-Zn-In) galvanic anodes as a source of DC current. Note that Al-Zn-In galvanic anodes require a certain level of chloride to activate and may not operate effectively when used in brackish waters
- impressed current CP, ie impressing a current onto the structure using an external DC source and inert or semi anodes, typically high silicon chromium cast (Fe-Si-Cr) anodes, mixed metal oxide coated titanium "MMO/Ti" anodes or platinum coated "platinised" titanium "Pt/Ti" anodes.

In order to provide adequate protection to all parts of the immersed structure, all piles should be bonded together to ensure electrical continuity. This can be best achieved on existing structures by welding the piles together. On new build structures where a concrete capping is used, it is desirable to weld each steel sheet pile to its neighbour before casting the cap.

Although galvanic and impressed current CP systems can be used in combination, this is unusual and it is normal for a single system to be applied. The two systems can, however, be used sequentially ie initial galvanic, followed by impressed current when the galvanic anodes reach the end of their life. The final choice of CP system for a given application is usually based on:

- an assessment of the design life required
- the mechanical robustness required

- the available funding for the initial installation
- the available provision of long term maintenance and operational surveillance expertise.

Advantages/disadvantages between the two types of CP systems, based on comparison tables in EN 12473:2000 and BS7361: Part 1 1991, are summarised as follows:

	Galvanic anodes	Impressed current CP
Environment	Use may be impracticable except in soils or waters with low resistivity. Aluminium anodes will not be appropriate for use in fresh water or some brackish waters.	Use is less restricted by resistivity of the soil or water but may get chlorine gas generation at the anode in seawater.
	There is some environmental concern about the dissolution of Zn galvanic anodes increasing the levels of zinc in enclosed waters and to a lesser extent in estuary areas (Van Damme, 2002; Matthiessen, 1999; Bird, 1995). Aluminium galvanic anodes are available with significantly reduced zinc levels (eg Al-Zn-In alloy anodes typically have a 4% Zn content). Possibility of bacteria (SRB) colonising anodes and reducing current output (Smart, 1997). Allowance should be made for self corrosion of anode materials by biogenic sulfides (King, 1989).	Chlorine generation has environmental implications but may be beneficial due to its known biocidal effect in killing bacteria.
Installation	They are relatively simple to install and additions may be made if the initial design is inadequate.	Design process is more complex. Impressed current systems are always more complicated than galvanic systems. The ease with which output may be adjusted may allow for unforeseen or changing conditions if the system capacity distribution is inadequate.
	The bolted or welded connections used to secure the anodes to the structure are inherently protected. There are no cable connections to anodes.	Requires high integrity of insulation on connections to the positive side of the rectifier which are in contact with soil or water, otherwise they will be severely corroded.
	Cannot be wrongly connected electrically.	Requires the polarity to be checked during commissioning and subsequent operation and maintenance because inadvertent electrical misconnection, with reversed polarity, would accelerate corrosion.
Power source	Independent of any source of electric power.	Requires a mains supply or other source of electric power. Solar power and wind generators may be considered for small applications where current demand is low.

	Provides protection to the structure on installation of the anodes or immersion of the structure with anodes pre-installed.	No protection until power becomes available. If power failure occurs, protective potentials normally remain for up to several hours if calcareous deposits are present on steel surfaces.
Anodes	They may be required at a large number of positions. Bulk and number of anodes may have weight loading and hydrodynamic (ie turbulence and drag) implications although these issues are not normally significant in any port or harbour application.	Lighter, more compact and fewer in number anodes. Anodes may be designed to minimise effect on water flow
Applicability	They are suitable for both coated and uncoated structures. The number and weight of anodes can be reduced by partial or complete coating of the structure.	Can be applied to all structures that can be protected by galvanic anodes and in environments where the higher electrical resistance of the electrolyte may prevent the use of galvanic anodes eg brackish or fresh water.
Control	Current output cannot be externally adjusted but a properly designed system will automatically vary the current output to match the current demand.	Controllable current output. Control usually automatic and may be continuous. Requires much more on-going monitoring, adjustment and maintenance by trained on-site staff than for galvanic anode systems. Liable to short term failure if inadequately monitored and maintained.
	By selection of suitable anode material, the structure cannot reach a potential that is sufficiently negative to damage coatings. All Zn and Al-Zn-In anodes are suitable for use with normal, high performance marine coatings.	Ineffective control or incorrect adjustment can lead to either overprotection (ie possible damage to coatings and/or hydrogen embrittlement at welds) or under-protection (ie structure not fully protected).
	Assuming a satisfactorily designed CP system, the level of protection is not much influenced by changes in surface areas due to tidal changes.	Variations in levels of protection if used in the galvanostatic (ie constant current) mode and, therefore, this method of control is unsuitable for high tidal range applications. Can reduce the effect of such tidal influences by application of coatings in the tidal zone or eliminate the difficulty by maintaining potentials automatically to within close limits (ie in "potentiostatic" mode) despite fluctuating areas of surface to be protected. The reliability of the permanent monitoring system, including reference electrodes, necessary for control in the "potentiostatic" mode can be poor.

Interaction	Unlikely to affect any neighbouring structures or berthed vessels because the current output of the individual anodes at any point is low	Effects on other structures situated near the anodes need to be assessed. Requires careful design and possible commissioning testing to minimise the risk of any "stray current" interaction with neighbouring structures, shipping at the quayside or any steel cables or pipelines lying on or within the mud zone.
Maintenance	Requires little or no attention other than periodic replacement of anodes over the design life of the structure. In practice, individual anodes may require replacement at different intervals of time on different parts of the structure.	Equipment designed for long life but regular inspection and maintenance required to keep electrical equipment (eg anodes, cables, reference electrodes and control system) in service. The design life of inert anodes can typically be up to 30 years. Reasonable to expect that the major electrical components (power supplies, anodes, reference electrodes, control and monitoring systems) will require replacement after 25 years (Wyatt, 1999). Those systems with buried anodes/groundbeds, which are less prone to mechanical disruption, are likely to be more reliable. Performance monitoring is more complex than with galvanic anode systems (see Section 3, Chapter 7). Function tests are required weekly by suitably trained on-site staff. Performance verification and system review are required at six monthly intervals by CP specialists.
Damage	Anodes are robust and not very susceptible to mechanical damage. Where a system comprises a large number of anodes, the loss of a few anodes has little effect on the system. Connection should be able to withstand any forces applied to the structure. Anode supports are also subject to mechanical damage due to impacts with shipping, debris or wave action.	Anodes lighter in construction and therefore less resistant to mechanical damage. Loss of anodes can be more critical to the effectiveness of a system.

2.1 Anodes

a) Galvanic anodes: The methods of attachment of the galvanic anodes to the structure need careful attention to ensure that they are firmly secured and can be replaced without difficulty. Anodes also need to be located to ensure an even current distribution across the entire structure. This is effectively achieved using stand-off anodes rather than flush mounted ones and is the best geometric arrangement to optimise current output. On steel sheet piled walls, the anodes are routinely attached within the inpan either by welding or bolting to brackets. It may be necessary to ensure electrical continuity of bolted brackets by tack welding. Anodes would be vulnerable to mechanical damage and a potential hazard to shipping activities if they were arranged as stand-off anodes proud of the outpans.

The majority of the galvanic anodes are ideally placed 1-2 m below MLWS level which corresponds to the critical area for ALWC. This positioning is also convenient for installation and maintenance and a deterrent against theft. For deepwater quays it may also be necessary to install galvanic anodes at the sea bed (or dredged) level in order to achieve adequate provision and distribution of current. Locating and replacing anodes submerged in mud can present difficulties although this is not often a problem with typical port and harbour structures.

Anodes which are intermittently immersed are without effect when the tide is out and will take some time to fully activate after exposure to the seawater. Anodes positioned in the tidal zone are also more susceptible to damage and would not be expected to contribute significantly to the protection of the structure if their immersion time is limited. Subject to adequate structural design, fixing of the anodes or their fixing brackets can be achieved prior to driving of the sheet piles.

Renewal of the anodes will need to be carried out by divers, possibly using underwater cutting and welding techniques; unless the anode to structure connections were designed with bolted brackets or similar. There is no necessity to remove consumed anodes unless they impede the installation of their replacements. Welded fixings which are continuously immersed in seawater ensure that they are fully protected from corrosion. It is also possible to design systems with extended steel cores to the anode in order that they can be installed without underwater works and welded to the sheet piles above low water. Such welding should be undertaken as near to low water as possible (certainly no higher than mid tide level) to ensure an acceptable level of corrosion protection to the welds. This extended steel core method facilitates anode replacement but the supports require rigorous design and placement to ensure that they are not susceptible to mechanical damage from impacts.

Bracelet anode assemblies are available for tubular piles and assemblies using clamped arrangements for other pile designs.

b) Impressed current CP anodes: Impressed current CP anodes are available and can be fitted into the in-pan areas of continuous steel sheet piled walls. Alternatively, they can be installed remote from the structure, for example on adjacent dolphins or other tubular elements or installed in a groundbed anode arrangement in the mud zone at some distance from the structure in order to optimise current distribution.

Due to the higher current output of impressed current anodes compared with galvanic anodes (typically 10 to 20 times) they may interact detrimentally with shipping or other nearby metallic structures or services (ie interaction or stray current effects). The risk is increased if they are positioned remote from the structure. Interaction testing procedures and acceptance criteria together with possible remedial action are given in the International and European Standards on CP (see Chapter 7, Section 3).

The anodes and cable connections may be susceptible to damage by shipping or dredging operations. In addition, any unstable bottom conditions may cause excessive coverage of the anodes which may reduce the current output if the electrical resistivity of the bottom sediments increases or as the distance to more conductive seawater increases.

Impressed current CP systems for ports and harbours typically utilise:

- anodes (typically MMo/Ti or Pt/Ti) integrally mounted on GRP based dielectric shields and mechanical carriers fitted as bracelet anodes onto tubular piles or on rigid frames into in-pans on steel piled walls
- anodes (typically MMo/Ti or Fe-Si-Cr) securely suspended from piles or decks or buried in sediment where shipping activity and any dropped debris will not damage the anodes and their cables.

Other anode arrangements specifically designed for low conductivity water applications to provide local protection in the low water zone and at the boundary between the submerged and mud zones (Sandberg 2002) include:

- an elongated anode strip fitted onto a wooden beam which functions as a mechanical support for the anode and achieves an optimal distance between the anode and piling and/or
- an anode wire placed inside a perforated plastic tube fitted onto the piling along the whole sea bed for protection just blow MWL or at the seabed respectively.

2.2 Cathodic protection design

It is recommended that the design of CP systems is carried out by suitably qualified and experienced personnel. The design of CP systems for steel in sea water is well documented in international codes and standards, notably in the European Standards EN 13174:2001 *Cathodic Protection for Harbour Installations, EN 12473:2000 General Principles of Cathodic Protection in Seawater*, EN 13509:2003 *Cathodic Protection Measurement Techniques* and DNV PR B401 1993 (specifically written for offshore only). British Standard BS 7361 is no longer a current standard. Guidance on the design criteria including current density, protection potentials, coating breakdown factors, etc are provided in these standards with respect to corrosion of steel in seawater.

None of the available standards give details of design criteria to be applied when an existing structure is suffering from ALWC. However on either new build or existing structures not suffering from ALWC, the standard design criteria should be sufficient to prevent the onset of ALWC. ALWC has not been noted on any structure where a well designed and fully maintained CP system has been installed at construction.

a) Protection criteria

Protection criteria have been collated from published literature on a variety of marine systems and, though not specifically derived for steel sheet piled structures, they are nevertheless applicable in principle to them. The potential of iron and steel in slowly moving aerobic seawater at ambient temperature is around -0.650 Volt (with respect to Ag/AgCl/seawater reference electrode). It is generally accepted that structures need to be held at a "Instant OFF" potential of -0.900 Volt (with respect to Ag/AgCl/seawater reference electrode) in order to protect against anaerobic corrosion by SRB (note that this potential is equivalent to -0.950 Volt with respect to Cu|CuSO4 reference electrode). The British Standard/European Standards and DNV documents specify -0.900 Volt (with respect to Ag/AgCl/seawater reference electrode) for anaerobic/SRB conditions. This electrochemical potential figure is referred to as "the protection potential" as it indicates the potential to which the structure should be polarised for full protection. Any design should be able to meet this protection criterion.

Both standards acknowledge the use of the -0.900 Volt protection potential criteria but discrepancies are present in the documents which influence the design procedures/calculations. DNV recommends the -0.900 V used for monitoring purposes but not in the design calculation; in page 28 of DNV RP there is the following comment "*...E = -0.8 V shall always be used for design calculations. This applies also to anaerobic environments with the protective potential (5.4) has been chosen to -0.900 Volt (Ag/AgCl/seawater). The latter potential is then applicable as a criterion for monitoring that adequate protection potential has been achieved, but does not affect the design calculations*". Conversely page 18 of EN 13174:2001 states: "*The commonly used net driving potential between an anode made of a typical aluminium or zinc based alloy and a polarised or coated structure at its minimum cathodic protection level (-0.800 Volt vs Ag/AgCl/seawater) is only 0.150 Volt to 0.300 Volt*". This could be interpreted that the use of a smaller driving voltage (and thus smaller current output delivered by the anode) is possible at the design stage. The latter seems to be the most appropriate for the ALWC applications as the protection potential criteria is included in the design procedure. The design provides anodes that have a higher length/radius (or equivalent radius) than would be the case if the DNV practice were followed. This is a key parameter for achieving lower anode resistance to sea and thus higher galvanic anode current output. The driving potential based on a -0.900 Volt polarised protection potential criteria is typically 0.150 Volt (-1.05 Volt anode operating potential less -0.900 Volt protected polarised potential) whereas DNV suggests the use of 0.250 Volt based on a polarised protection potential of -0.800 Volt. Therefore, at an equivalent anode mass (and thus life) an anode based on the DNV procedures would deliver less current than its counterpart based on the above mentioned procedure.

b) Design criteria

To arrive at this protection potential a certain current should be applied to polarise the structure sufficiently. This current applied over an exposed surface to be protected is quoted as a unit of current density (usually mA/m^2). The current densities required for coated and uncoated steel in moving and stagnant seawater and in soil areas vary within large ranges depending on local conditions.

It is essential that the current provision to achieve the above protection criterion under the full range of environmental conditions (eg reduced salinity due to fresh water run off in estuarine conditions or due to temporarily high current requirements caused by high flow rates or storm conditions) is available at all locations over the structure, otherwise only partial protection will be achieved at these areas.

None of the available standards and/or recommended practices has specific guidance on current density requirements for existing structures suffering from ALWC and the values quoted may not be sufficient to arrest the ALWC. This is an area where further research/investigation is required.

For new build structures and existing structures not presently suffering from ALWC guidance on current density should be obtained from European Standard EN13174:2001 which surmises in Annex A "*Guidance for current requirements for cathodic protection of harbour installations*" but qualifies the information by stating "*in the absence of any other documented experience, the following values may be considered*". Therefore, for structures with established ALWC the upper figures, at least in respect of the Initial Values, should be used until further clarification is obtained by research and/or experience. The quoted design current densities for the protection of bare steel in seawater are tabulated below:

Situation	**Current density (mA/m²)**					
	Initial value		**Maintenance value**		**Repolarisation value**	
	in poorly aerated waters	in well aerated waters	in poorly aerated waters	in well aerated waters	in poorly aerated waters	in well aerated waters
Static or semi-static conditions with tidal flow less than 0.5 m/s	80 to 100	120 to 150	50 to 65	65 to 80	60 to 80	80 to 100
Static or semi-static conditions with tidal flow more than 0.5 m/s	120 to 150	170 to 200	60 to 80	80 to 100	80 to 100	100 to 130

Design current densities for the protection of bare steel in saline mud are as follows:

	Current density (mA/m²)		
	Initial value	**Maintenance value**	**Repolarisation value**
All types of structures	25	20	20

c) Other design information

Other than the current required, the following information is also needed to design a CP system:

- resistivity of electrolyte in which the anode will operate at the appropriate temperatures
- quality and breakdown of any coating installed

- electrical continuity of structure
- anode type and distribution
- monitoring facilities
- pile diameter for tubular piles, length or type of pile for steel sheet piling
- pile cut off level
- tidal levels
- sea bed (dredged or otherwise)
- toe level (or founding level)
- information on fendering
- general layout of facility.

Additional design information to be considered for galvanic anodes includes:

- anode geometry (shape and stand-off)
- anode core design to deliver Utilisation Factor (the proportion of the active material that can be effectively utilised) and the end of life current of the consumed anode in order to meet the Repolarisation Current Density
- anode dimensions and mass
- anode fixings for long term durability and electrical continuity
- for Al-Zn-In anodes the salinity of the electrolyte.

Additional design information to be considered for impressed current CP systems includes:

- availability and location of DC power source
- safe and secure location for CP transformer rectifiers, secure and robust cable management systems
- distribution of current and provision of additional current density figures (over those recommended above) to accommodate non-uniform distribution
- permanent potential monitoring devices to allow automatic potentiostatic control in locations with significant tidal range or variation in water flows or salinity
- cable runs and distribution boxes
- overprotection; possible damage to coatings or high strength steels
- stray current.

3.0 Concrete encasement/jacketing

Another method of protecting steel on both steel sheet piled walls and skeletal structures is to encase or jacket it in concrete. The concrete acts as a barrier and the alkaline environment created by it passivates the steel. Both precast concrete elements (Sandberg, 2002; Vesth-Harven, 2002) and underwater concrete, cast in-situ (see Chapter 6, Section 6.2) have been used. Concrete is a brittle material and, therefore, concrete jackets are also susceptible to cracking as a result of excessive steel movement, impact from vessels or debris striking the structure. Therefore, the method is only suitable for structures where the working stresses in the concrete are held to reasonable values.

Depending on the thickness and quality of the concrete used, corrosion problems on the concrete covered steel surface and/or reinforcement bars (if used) can also occur over time due to:

- build up of aggressive salts (eg chlorides) resulting from permeation of seawater into the concrete or as a result of cracking
- breakdown of passivity from carbonation of the concrete above water.

Chloride penetration or carbonation and the subsequent corrosion of the steel may eventually lead to spalling of the concrete due to a "rust jacking" mechanism (ie stresses generated due to the volume increase associated with oxide formation). The use of pre-coated reinforcement bars and corrosion inhibitors in the concrete mix (Perry, 2001) have been used in an attempt to reduce the risk of such corrosion occurring. Consideration should also be given to the possibility of bacterial attack on the concrete itself (Emami, 1998) along with other possible chemical attack as may be appropriate to the specific location.

To effectively protect steel from corrosion the concrete should:

- be of good quality
- be of high durability
- be relatively impermeable
- be free of significant cracks
- have good bonding characteristics
- be properly placed, compacted and cured
- be of adequate cover over the reinforcement
- have high strength (occurs as a result of the other features).

Concrete used in tidal and splash zones should be designed for durability with low water/cement ratio, cement replacement with silica fume, ggbs (ground granulated blast furnace slag) or similar materials to reduce chloride diffusion, be well compacted and well cured.

Several standards are available which give general guidance on concrete mixes for a variety of environments including maritime (eg BS 8500 which was issued in December 2002 and is complementary to BSEN206 Parts 1 and 2). In practice, the steel should be initially blast cleaned to remove surface deposits in order to provide a good mechanical bond for the concrete. Also, shear studs are normally welded onto the structure to provide a connection for the reinforcement steel mesh. A concrete cover in the range of 50 mm to 100 mm over all the steel (both structural and reinforcement) should be provided depending on the application. Ideally the number of joints should also be minimised and special attention should be given to the seals at construction joints.

In the past, concrete encasement has been used to protect steel in the atmospheric, splash and tidal zones. It has rarely found application in protecting steel in the submerged zone due to the practical difficulties and costs. To be effective against ALWC, the concrete facing should ideally extend down to bed level in order to minimise the risk of additional corrosion cells being set up on the structure. This is because in the case of partial jacketing, the steel within the concrete would be expected to assume a noble potential and would, therefore, be liable to serve as a cathode in a macro-corrosion cell with the remaining unprotected submerged steel becoming the anode (AISI, 1981; Wyatt 1999). This has been reported to have occurred in one application when a concrete lining, originally for protecting the splash zone, was lowered to the low water level and the corrosion rate just below the lining was found to have increased (Zen, 1985). However, it has been argued (Sandberg, 2002) that this accelerated attack does not occur to any significant extent. This is because the embedded steel in water saturated concrete is said to become less noble than the bare steel in seawater due to the low diffusion rate of oxygen in saturated concrete. This argument has been used to justify terminating concrete facings 1-2 m below MLW level since the piling directly below the concrete is said not to be negatively influenced. Another reference (Steel, 2002) also describes a recent application where U piles were protected with a concrete facing to give a design life 80 years. The facing was 100mm thick nominally reinforced concrete connected via shear studs to the steel wall. The concrete was placed using two sections of shuttering the lowest of which was installed using divers. The concrete face covered both the splash zone down to the LAT zone. Others, however, using this partial concrete facing approach have provided galvanic anodes within the submerged zone to protect against any increased corrosion attack in this area (eg Vesth-Harven, 2002).

4.0 Combinations

There are two equally valid ways of looking at organic coatings used in combination with CP for corrosion protection. From a coating engineer's viewpoint, the coating is the primary protection while the CP's function is simply to protect the substrate steel at inevitable "holidays" or damaged areas within the coating. From a CP engineer's viewpoint, the CP provides the primary protection whilst the coating is applied simply to reduce the current requirements from the CP system. Also, in offshore applications, there are two schools of thought as to whether CP and coatings should be used in combination for protecting steel surfaces which are submerged all the time (eg platform jacket legs). Some advocate the use of high performance coatings in combination with CP to reduce the current requirements and, in the case of galvanic anode CP, to reduce the number and weight of anodes. Others prefer to leave steel bare in the entire submerged and soil zones because it removes the risk of possible shielding affects from protective current at areas of disbonded coatings and it leaves all surfaces clear for underwater inspection and NDT work that may be required.

If organic coatings are used, good surface preparation, removal of surface contamination (eg salts) and coating application is essential to reduce the risk of osmotic blistering. If impressed current CP systems are used, good monitoring and control of current output and the resultant protective potentials are required to minimise the risk of overprotection giving coating disbondment or hydrogen embrittlement of high strength steels or sensitive weld areas.

In the case of steel structures subject to tidal influence, the primary purpose of the coating is to protect the steel where CP does not act (ie mid tidal zone and above). Depending on the actual mechanism of ALWC (see Chapter 2, Section 2.3.3), the presence of a coating to steel surfaces in the tidal zone may also reduce the rate of ALWC by reducing the cathodic area of the macro corrosion cell operating between this area and the low water zone (see Appendix 2, Section 2.0). For large surface areas of exposed steel, a coating system can dramatically reduce the current required for complete protection. As a result, there will be an expected significant reduction in the number and/or size of galvanic anodes or the impressed current demand.

The use of a coating as an adjunct to CP is also advantageous in complex and dense geometries where distribution of CP current is difficult or where high sea water velocities occur. The current density for protection of bare steel is known to increase with increasing water velocity; so the use of CP and coatings lower the current density required for protection of coated steel surfaces and also allows better current distribution compared to bare steel.

Overprotection with impressed current CP is to be avoided. Once the surface "Instant OFF", IR error free potential becomes more negative than around -1.200 Volt (with respect to Ag/AgCl/seawater) a region of overprotection is entered and there is a risk of coating disbondment. Also associated with overprotection is alkali generation which can be deleterious for some coatings. Ideally, "Instant OFF" potentials should be maintained within the range of -0.900 Volt to -1.150 Volt (with respect to Ag/AgCl/seawater). Coatings should be fully tested for compatibility with CP (ie cathodic disbondment testing). Written confirmation from both the CP system designer and the coating manufacturer should be obtained to demonstrate compatibility. Cathodic disbondment tests to BS3900-F10, or equal standard, can provide corroborative data.

To provide adequate protection and be compatible with CP, a protective coating should exhibit the following properties:

- maintenance of high dielectric properties for coatings is desirable to resist electron passage (ie to be a good electrical insulator)
- coatings should be chemically resistant, particularly to saponification by the alkali produced at the cathodic steel surface and specific pollutants that may be present at the particular site
- coatings should have a low moisture absorption and transference rate in order to prevent hydrogen blistering and good resistance to osmotic blistering
- optimum coating thickness should be used to maintain the barrier effect for the cathodic current
- possess good abrasion resistance, hardness and shock resistance.
- exhibit excellent adhesion to the base metal
- exhibit low susceptibility to underfilm corrosion in areas that become damaged
- be easily applied and easily repaired.

The following coating systems have been proposed (Cunningham, 1998) for use with the CP of steel sheet piling.

System number	**Use**	**System**	**Dry film thickness (μm)**
1	System suitable for immersion with or without CP.	**High-solids epoxy:** • direct-to-metal epoxy • direct-to-metal epoxy.	 150 150
2	Heavy-duty, reinforced system suitable for immersion with or without CP. Improved abrasion and wear resistance.	**Glass-flake-reinforced high-solids epoxy*:** • epoxy primer • epoxy glass flake.	 75 400
3	Single-coat, solvent-free system suitable for immersion with or without CP.	**Solvent-free phenolic epoxy:** • phenolic epoxy.	400

* Note that this system should be overcoated with a polyurethane in high UV areas adjacent to water to prevent rapid deterioration (ie "chalking") of the epoxy coating.

Small areas of damage to these systems can be repaired by application of a surface tolerant epoxy following suitable surface preparation. Larger areas of Systems 2 and 3 should be repaired with the original coating materials, as required.

A11 ALWC CASE STUDIES

CASE STUDY 1

Atlantic Wharf, Aberdeen Harbour: Use of welded strengthening plates and protection using high build epoxy coating.

- **type of structure and design life:** Larssen 5 **Steel Sheet/Box Pile Wall**, original design life unknown, **constructed 1948**. Original corrosion protection unknown, no evidence of any
- **how was residual life predicted**: Observation showed that unless remedial measures were affected promptly there would be no residual life. This berth had just had a new deck and rail mounted cargo crane installed above so it was critical **to rescue it from reconstruction**
- **ALWC early/intermediate/advanced and how detected: ALWC advanced**, originally detected visually from **boat survey**, closer low tide inspection revealed severe erosion/pitting over large areas of pile face and in some cases **holes and loss of fill**. However, this **structure was 50 years old** at the time
- **approach taken and why:** At the time coating of piles was considered as adequate protection. In this case the piles were overhung by a deep concrete cope above (0.6 m) and used to be fendered in timber. They were set back and therefore **protected from berthing damage**. A particular brand of **epoxy coating** tailor made for this application was considered and after some references and site visits this coating system was adopted. However **repairs** to many outpans were also necessary and it was decided to **U plate every outpan whether or not it needed it** – this made quantifying straight forward. U plates were fitted by **intermittent weld** with **edges sealed** with a resin compound and then a non-shrink **grout pumped into the void** between the plate and original pile face. On holed piles the **holes needed to be patched locally first**.
- **repairs done, application, and cost:** 167 outpans plated over a height of 2.2 m and berth length of 143 m centred on LAT
- **observations/lessons learned in repair stage:** The importance of the contractor to have an efficient and **effective strategy to deal with holed piles** cannot be overstated. Holes that cannot be detected otherwise become plainly obvious by passage of ground water from behind the wall once inside the dry setting and after blasting clean the steel piles. Water running down the steel piles is a great hindrance to welding and in this case to coating. Likewise the importance of a dry setting that is tight is also vital for similar reasons. The more wobbly the line of sheet piles the more difficult it is to achieve **good dry setting**, unless the setting itself is quite sophisticated in terms of adapting
- **protection methods single/hybrid considered:** Entire exposed sheet-pile faces over a height of 2.5 m centred on LAT blasted to SA21/2 then coated with 400 μm epoxy in one film. **Access** gained by way of **"dry setting installation"** covering 560 sqm. The **remainder of piles** from the bottom edge of coating to bed had a very hard natural coating of crustaceans/calcium about 5 mm thick which, **left undisturbed**, would continue to protect the piles
- **reasons for final choice:** Bids for similar processes were also submitted but this method won **on basis of cost**
- **relative costs of protection methods used:** The repair and protection cost on this berth finally worked out at **£2000 per l m fully inclusive in 1999/2000**. It is difficult to calculate a split as the considerable method related charges were common to both repair and protection but it works out at approximately **£1400 per l m repair and**

£600 per l m coating over a 2.5 m height. The works took **21 weeks to complete**

- **ALWC prevented?:** Yes
- **repair/protection design life:** 30 years
- **observations/lessons learned:** In addition to above, the **importance of a dry substrate for coating** is also critical. Therefore, leaky clutches, cope above or dry setting is not good. Any type of moisture and the coating will not adhere, even though with this product was satisfactory to immerse immediately after application. Furthermore, on this berth holing tended to be on the webs of the piles rather than the outpans, although out-pan faces were also severely dished on many of the piles. This was **the first time this coating product had been used in UK** and as a consequence environmental authorities needed strong convincing that it was harmless to the environment.

CASE STUDY 2

Trinity Quay, Aberdeen Harbour: Use of welded strengthening plates and protection using galvanic anodes.

- **type of structure and design life:** Frodingham 4N **sheet pile wall**, original design life unknown, **constructed 1974**. There was no record of original corrosion protection of any kind, although a few anodes appear to have been installed at random sometime after construction
- **how was residual life predicted:** It was clear from the **holing** that action was needed immediately
- **ALWC early/intermediate/advanced and how detected: Advanced ALWC** at centre of berth over 20 m length approx, severe holing in webs towards outer corners from LAT to MLWS. This location is approx. mid-ships of a berthed vessel and **stray current corrosion is suspected as a contributing factor**. Detected by observation during low water **boat survey**
- **approach taken and why:** This sheet pile wall is tidal both sides, it would not be possible therefore to place a dry setting at holed areas to do repairs. A thorough **survey at low water and by diver to bed** was carried out prior and it was possible to chart the exact condition of the piles and to quantify areas requiring repair. There was also **some advanced corrosion** at bed towards the corner of the berth **from propeller wash and scour**
- **repairs done, application, cost: Profiled steel plates welded** all round over holed/thin areas (50 per cent or less original steel thickness), applied **by diver**. **£600 per 1.2 m high side-pan patch plate fitted**
- **observations/lessons learned in repair stage:** Importance of **accurate** in-situ pile **profile measurement for repair plates**. The Z shaped patch plates had to be bent accurately for a snug fit. A welder who can dive produces a quality result in a competitive time-scale
- **protection methods single/hybrid considered:** Coatings ruled out because dry setting by limpet not possible, and the steel pile outpan faces are particularly exposed to berthing impact and abrasion (rubbing). **Cathodic protection with anodes** placed on the in-pans behind the berthing line was selected
- **reasons for final choice:** This berth had anodes fitted previously at random locations and there was no ALWC evident at these locations. It was clear they had performed so the same strategy was applied but along the entire berth
- **relative costs of protection methods used: £500 per linear metre of berth for 130kg Aluminium anodes every in-pan**
- **ALWC prevented?:** Yes
- **repair/protection design life:** 20 years
- **observations/lessons learned:** The importance of **engaging corrosion engineers** to

map the corrosion status of the structure allowing optimum design of the most appropriate cathodic protection solution.

CASE STUDY 3: **Upper Quay, Aberdeen Harbour: Protection using galvanic anodes.**

- **type of structure and design life:** Open quay structure bearing on a matrix of 356 × 368 × 174 kg/m **UBPs**, original design life unknown, **constructed 1976**. Original corrosion protection consisted of 200 μm bitumastic biocide coating, long since depleted
- **how was residual life predicted:** After **measuring typical corrosion losses** at the most severe areas (at low water), **structural calculations** were carried out to verify the integrity of the piles. **Stopping corrosion** immediately would suffice **without repair** being necessary
- **ALWC early/intermediate/advanced and how detected: Advanced ALWC** on outer flanges of outer (quayside) row of piles only. Other deeper rows of piles in good condition. Detected by observation during low water **boat survey**, followed up by detailed residual thickness measurement
- **approach taken and why:** See below
- **repairs done, application, cost:** No repairs necessary to affected piles. The piles affected happened to be doubled up in pairs to carry fender support brackets jointly, therefore an outer flange steel loss of 50 per cent was still not critical, although had to be curbed
- **observations/lessons learned in repair stage:** N/A
- **protection methods single/hybrid considered:** Although such piles lend themselves to coatings (no risk of scour, mechanical abrasion) coatings ruled out because of **awkward access and safety considerations**. **Diver application of anodes** to every immersed pile (128 no.) the simplest and most cost effective corrosion arrestment method under the circumstances
- **reasons for final choice:** An adjacent and similar berth had anodes fitted previously and there was no ALWC evident at these locations. It was clear they had performed so the same strategy was applied to this berth
- **relative costs of protection methods used: £600 approx. per pile for single 178kg Aluminium anodes every pile (fitted)**
- **ALWC prevented?:** Yes
- **repair/protection design life:** 20 years
- **observations/lessons learned:** The importance of **engaging corrosion engineers** to map the corrosion status of the structure allowing optimum design of the most appropriate cathodic protection solution. Importance of **accurate bed level measurement** at each pile to ensure adequate submerged height for anode installation.

CASE STUDY 4: **Bulk Handling Jetty, Western Australia; Protection of jetty piles using a reinforced concrete infill technique in conjunction with a wrapping system.**

- **type of structure and design life:** T shaped **jetty** added as an extension in **1977** comprising 900 mm thick reinforced concrete slab supported on an arrangement of **tubular piles** each 675 mm diameter with a 20 mm wall thickness. Original corrosion protection was by a wrapping system at the low water mark in place for 10 years
- **how was residual life predicted:** Unknown
- **ALWC early/intermediate/advanced and how detected: Advanced ALWC**; when the first wrapping system was applied many of the piles were found to be holed at the low water mark underneath heavy marine growth

- **approach taken and why:** The existing wrapping system reached the end of its useful life. A renewal programme using a **petrolatum paste based wrapping system** applied over a height of 2 m extending to 0.5 m below the low water mark protected by a circumferential jacket. Above this was applied a double-wrap petrolatum tape based system up to the soffit at +4.0 m datum along a single line of 62 piles. The lower part of the combined system consisted of a jacket wrapped over a paste pulled together and tightened along a vertical seam by divers, while the upper part consisted of a spiral wrap applied in the dry
- **repairs done, application: A concrete pile infill technique** was initially used for plugging and filling the piles for strengthening as follows:
  a) If a pile was holed, then divers first fixed a temporary plate.
  b) A 150 mm diameter hole was then cored through the deck into the centre of each pile below.
  c) For holed piles water was then removed to a level 2-3 m below the hole. If not holed, each pile was filled with water to similar level.
  d) A two part liquid urethane polymer was then mixed and dropped through the deck hole onto the water surface inside each pile to form a plug. This procedure was repeated to build a high density plug approximately 1m deep. Sufficient concrete was then poured onto the urethane as a base for the reinforcing steel.
  e) The following day a prefabricated expanding reinforcement cage was dropped through the deck hole and fill with 32MPa concrete to deck level.

CASE STUDY 5: Tug berth, Eastern Australia: Protection of sheet pile wall using replacement sheet piling.

- **type of structure and design life:** Original sheet pile wall 97 m long consisting of SP108 U section steel sheet piling coated from the underside of cope down to 1 m below low water. Scour protection installed at bed level consisting of 225 mm thick concrete blocks laid on a geo-textile membrane
- **how was residual life predicted:** Unknown
- **ALWC early/intermediate/advanced and how detected: Advanced ALWC** sheet piles were discovered to be heavily corroded from the low water mark throughout the low water and submerged zones and down to mud level. A video of the dive survey was viewed which revealed three typical types of corrosion in three bands, viz:
  a) Very localised, active corrosion in the low water zone on the outpans over the face and sides resulting in pits 20 to 25 mm diameter and 8 to10 mm deep, sometimes holed through.
  b) Localised, active corrosion in the mid corrosion zone over relatively larger areas on outer faces towards corners, in some cases resulting in holes up to 50mm wide and 100 mm long.
  c) Splits or ruptures on side pans immediately at the clutch over a height of about 1.5m from bed level. Deterioration looked old with no sign of active corrosion.

- **approach taken and why:** A cost-benefit analyses concluded that replacement sheet piling immediately to the front of the existing was the best solution
- repairs done, application: A design-and-construct proposal was accepted that consisted of installation of new AZ36 Z section sheet piles approximately 1m in front of the existing wall tied in to the existing tie rods and top cope beam and backfilled with sand, taking advantage in the new design to deepen the berth by approximately 2 m. The new sheet piles were **pre-coated** over the top 8 m of the seaward face with a two component, high build, high solids polyamide-cured

surface tolerant epoxy mastic coating. To protect the existing structure along with the rear of the new sheet piles from corrosion, **cathodic protection** chambers were formed between the two walls, accessible via covers on the quay surface and a serviceable **impressed current** system installed.

CASE STUDY 6: **Port of Ipswich: Use of increased section thickness in the low water zone, by attachment of profiled steel plates in conjunction with coating applied to surfaces above low water zone.**

- **type of structure and design life**: New quay wall. The works were procured under a design and construct contract with a specified design life and specified life of ALWC protection. Design life unknown
- **approach taken and why:** There was evidence of ALWC in the Port on both tubular piles and sheet piles, although not as aggressive as has been noted at other Ports. Therefore, it was important that, when the new quay wall was proposed, full account of the potential effects of ALWC was considered. Options for ALWC protection proposed at tender stage included a fully painted solution, cathodic protection (sacrificial) and sacrificial steel plates at LAT level with paint above. All options were offered with the same design life but it was considered that the fully painted solution could suffer damage during driving and also from the subsequent dredging operation. Galvanic anodes would need to be installed after the dredging was complete, which may have cost and programming implications. Therefore, it was decided to install a steel plate system.

  Steel piles were first driven under dry conditions. Profiled steel plates were then welded onto the structure in the ALWC susceptible zone. Coating was applied to steel surfaces above the low water zone and the weldments used to secure the plates. The aim of coating application to these areas is to significantly reduce the cathodic areas of the structure thereby reducing the macro cell effects and resulting corrosion rate on exposed steel in the low water zone. Corrosion of the steel plates to provide protective current under immersed conditions to any damaged areas of coating.
- **application cost:** unknown
- **ALWC prevented or reduced?:** Too soon to make any judgement on strategy used. The quay wall will be monitored regularly and if and when necessary the ALWC protection can be repaired or supplemented with sacrificial anodes.

TABLE OF CATHODIC PROTECTION CASE STUDIES*

Project location	Year of installation	Pile type(s)	New or existing structure?	Coating?	CP type	Design life (years)	Sub-merged length (m)	Buried length (m)	Total surface area (m^2)	Cost/m^2 (£)
NE Scotland	2002	Larssen 20W and 32W	Existing	100% broken down	Galvanic	25	14	8	9000	20
NW Scotland	2003	Froddingham 4 and Larssen 6 and tubular piles	Existing	100% broken down	Galvanic	20	7	8	16000	20
W Scotland	2003	Larssen 32W	Existing	None	Galvanic	20	6	5	4500	30
W Ireland	2003	Combi LX32 + tubular	New	New (10% damage)	Impressed Current	25	14	11	15500	10
E Ireland	2001	Larssen LX38	New	None	Galvanic	15	10	9	7500	15
NE England	2004	Combi LX16 + tubular	New	None	Impressed current	20	11	9	10000	15
NE England	2003	Froddingham SW-1A	Existing	None	Galvanic	20	12	2	3000	35
S England	2003	Froddingham	Existing	None	Galvanic	25	5	7	2000	30
SW England	2001	Combi Larssen 16W + tubular	Existing	100% broken down	Galvanic	20	5	5	3000	30

Notes:

(1) Cost includes design and installation of system.

(2) For impressed current the cost excludes annual monitoring and electricity supply charges.

*Provided by Corrosion Control Services Limited, Telford

Bibliography

A Standards/guidance documents/booklets

BRITISH STEEL PILING SALES AND TECHNICAL SERVICES
"Accelerated Low Water Corrosion of Steel Sheet Piling"
Pub. No. P128 Sept. 1996

BS6349: PART 1: 2000
"Code of Practice for Maritime Structures – General criteria"

COASTAL DEVELOPMENT INSTITUTE OF TECHNOLOGY
"Manual for corrosion control and repair of Port Steel structures"
December 1996, 1-465

CORUS CONSTRUCTION & INDUSTRIAL
"Durability and protection of steel piling in temperate climates"
March 2003

Hodgson M
"Manual for the repairing of sheet piled and other quays using limpet dams"
John Martin Construction Ltd.

ICE MARITIME BOARD
"Concentrated Corrosion on marine steel structures",
2000

NACE STANDARD RP0272-72 ITEM NO 53009
"Direct calculation of economic appraisals of corrosion control measures" 1972

NACE
"Forms of Corrosion Recognition and Prevention"
1981, 1-116

OCDI
"Corrosion Protection and Repair Manual for Port and Harbour Steel Structures"
The Overseas Coastal Area Development Institute of Japan, 1998

PIANC PTC II WORKING GROUP 17
"Inspection, maintenance and repair of maritime structures exposed to material degradation caused by salt water environments"
Supplement to Bulletin No 71 1990

THE INSTITUTION OF STRUCTURAL ENGINEERS
"Guide to Inspection of Underwater Structures"
October 2001

B Reference books

BRITISH STEEL
"Piling Handbook"
7th Edition 1997

LAQUE F L
"Marine Corrosion"
John Wiley & Sons, 1975

SHRIER L (ED)
"Corrosion"
Newnes Butterworth, London, 1978

THORESEN
"Port Designer's Handbook: Recommendations and Guidelines"
Thomas Telford, 2003

UHLIG H H
"The Corrosion Handbook"
John Wiley & Sons, Inc. 1948, 1-1188

C Conferences/seminars

"***Corrosion of Sheet Piles – Can We Slow it Down?***" ICE/PIANC Half-day Meeting at ICE, London, 1994

"***Accelerated Low Water Corrosion – A Matter of National Importance***" ICE Maritime Board, One-day Meeting at ICE, London, October 1998. Papers comprised:

1. Carmichael J et al "ALWC – An Introduction", Arch Henderson & Partners.
2. Pritchard A "Wider experience of bio-corrosion", AEA Technology.
3. Beech I "Introduction to the theory of biocorrosion", University of Portsmouth.
4. Gubner R "Statistical assessment of the ALWC – risk in tidal waters", University of Portsmouth.
5. Marsh E "Protection methods of steel sheet piling – UK experience", British Steel.
6. Reinhard K"German experience", Hoesch Spundwand und Profil.
7. Sovang T "Underwater Corrosion of Sheet Piling", Acotec N V.
8. Hodgson M "Rectification, restoration, recovery", John Martin Construction Ltd.
9. Wright P "A designer's view", Posford Duvivier.
10. Beech I "Further research, publications and funding", University of Portsmouth.
11. Roberts J "The extent of the problem (a confidential reporting survey?)", Allott & Lomax.
12. Shelley T "ALWC Experience in Piled Structures", Port of Felixstowe.
13. Anon. "BS6349 redraft".

"***Accelerated Low Water Corrosion***" – ICE Maritime Board/East of Scotland Association/INA, half-day meeting at Herriot-Watt University, Edinburgh September 1999. Papers comprised:

1. Carmichael J et al "ALWC – An Introduction", Arch Henderson & Partners
2. Shelley T "ALWC Experience in Piled Structures" Port of Felixstowe
3. Wyatt B S "Cathodic Protection: A Solution for Accelerated Low Water Corrosion" Corrosion Control Services
4. Hodgson M "Limpet Dams and Repair" John Martin Construction Ltd.

5. Christie J "Steel Pile Corrosion Survey Model – Inspection Procedures for Accelerated Corrosion on Marine Steel Piles" Aberdeen Harbour.
6. Shaw M "Diving Inspections and Corrosion Below Lowest Astronomical Tide" Kirk McClure Morton.

"***Accelerated Low Water Corrosion 2002***" – Millennium Conferences 2-day meeting including technical visit, Marriott Hotel, Copenhagen, Denmark, April 2002. Papers comprised:

1. Chakraborty S "Corrosion – an overview and an introspection at low water", Haldia Dock Complex, India.
2. Graff M "Accelerated Low Water Corrosion (ALWC) and Microbial Induced Corrosion (MIC): Two Sides of the Same Coin?", Dr Graff und Partner.
3. Sandberg B "Cathodic Protection of Steel Sheet Piling in Brackish Water", Swedish Corrosion Institute, Sweden.
4. Shelley T "ALWC Survey of UK installations, 2001", Shelley Consulting Services.
5. L van Damme "Belgian system against ALWC" Ministry of the Flemish Community, Belgium.
6. Vest-Hansen "Routine inspection and protection of steel sheet piles in the Port of Copenhagen" Copenhagen Port Authority, Denmark.
7. Steel G "Accelerated low water corrosion experiences from The Port of Felixstowe and Port of Southampton", Associated British Ports.
8. Hodgson M "Stick with limpets – they will always get you down!", John Martin Construction.
9. Ligertwood P, James R "Corrosion is inevitable but it can be controlled – Corrosion Management Systems" CorrOcean Ltd.
10. Wyatt B "Cathodic Protection: A solution for accelerated low water corrosion", Corrosion Control Services Ltd.
11. Marshall D, Breakell J E "The development of an innovative treatment for dealing with Accelerated Low Water (ALWC) problems in Ports and Harbours", BAC Corrosion Control Ltd./Mott MacDonald Ltd.
12. Sonck W "Accelerated Low Water Corrosion", Acotec NV, Belgium.

"***Accelerated Low Water Corrosion***" – Session 2B, PIANC 30th International Congress, Sydney, Australia, September 2002. Papers included:

1. Christie J B "Accelerated Low water Corrosion – A Practitioner's Perspective" Aberdeen Harbour Board.
2. Kumar A, Stephenson L "Accelerated Low water Corrosion of Steel pilings In Seawater" US Army Corps of Engineers.
3. L van Damme, W Vrelust "Continuous Fight Against ALWC: An Evaluation" Ministry of the Flemish Community.
4. Beitelman A "Application Parameters For Arc Sprayed Metalised Coatings"US Army Corps of Engineers.

D Published research studies

COLENBRANDER W et alET AL
"*Life Cycle Management of Port Structures General Principles*"
Report of PIANC Working Group 31 ISBN 2-87223-103-X

COMMISSION OF THE EUROPEAN COMMUNITIES TECHNICAL STEEL RESEARCH
"Steel piling corrosion in marine environments: a survey"
Report EUR 7430

MORELY J, BRUCE D W
"Survey of steel piling performance in marine environments"
ECSC Report No. T/RS/1195/21/82/C, 1983

WIJNGAARD B H
"Steel piling corrosion in marine environments: a survey"
ECSC Report EUR 7430 1982

E Thesis/dissertations

Gubner R J
"Biofilms and accelerated low water corrosion of carbon steel piling in tidal waters"
Ph.D thesis University of Portsmouth 1998

M STENNULAT
"Accelerated low water corrosion; current methods for the protection and repair of sheet pile walls"
B.Eng. Dissertation, University of Greenwich, 1999/2000

F Technical papers/references

ANON
"Concentrated Corrosion - A New Phenomenon on Berth and Jetties?"
Port Engineering Corrosion, July/August 2001

ANON
"The Corrosive influence"
Port Development International, March 1999

BEECH I
"Introduction to the Theory of Bio-Corrosion"
University of Portsmouth, 1998

BEITELMAN A
"Application Parameters for Arc Spray Metallized Coatings"
PIANC 2002 30th International Navigation Congress, Sydney, September 2002

BENSON J AND EDYVEAN R G J
"Biologically Influenced Hydrogen Entry into Steel in Seawater"
Corrosion Management, May/June 1999

BREAKELL J E
"Repair of accelerated low water corrosion on marine SSP walls"
The Steel Construction Institute, 15th Steel Piling Group Meeting, London 2000

BRITISH PORTS ASSOCIATION
"A Guide To The Choice And Application Of Painted Protective Coatings", 1985

BRITISH STEEL CORPORATION
"The Corrosion and Protection of Steel Piling"
No T/CS/1115/1/79/C 12, November 1979

BRITISH STEEL GENERAL STEELS
"Coatings for Steel Pilings"
BSC Piling, March 1990

CHRISTE J B
"Concentrated corrosion on berths and jetties – accelerated low water corrosion a port engineer's perspective"
79th Annual Meeting and International Coatings Technology Conference, Edinburgh, September 2001

CHRISTIE J B
"Concentrated Corrosion on Berths and Jetties – Accelerated Low Water Corrosion"
30th PIANC Congress and ALWC Technical Tour, Australia, September 2002

CHRISTIE J B
"Repair of steel harbour piles damaged by concentrated corrosion"

DEXTER S C AND SHIANG-HO LIN
"Effect of Marine Biofilms on Cathodic Protection"
NACE, 1992

EDYVEAN R G J, MAINES A D ET AL
"Interactions between Cathodic Protection and Bacterial Settlement on Steel in Seawater"
International Biodeterioration and Biodegradation, 1992

GEHRKE T, KARIUS R AND SAND W
"Low-Water Corrosion on Steel Piling Structures caused by MIC"
Eurocorr 99

JEFFREY R AND MELCHERS R E
"Bacteriological influence in the development of iron sulphide species in marine immersion environments"
Corrosion Science, 45, 2003, 693-714

KAZUAKI Z
"Corrosion and corrosion control of port facilities"
Corrosion Engineering, Vol. 34, No. 3, 1985, 184-189

KAZUAKI Z
"Corrosion and Corrosion Control of Steel Structures under Marine Environment"
4th Asian-Pacific Corrosion Control Conference Material Conservation & Corrosion Control, Vol. 1, Tokyo, Japan, May 26-31, 1985

KAZUAKI Z ET AL
"Applicability of CP to concentrated corrosion"
Report of the Port and Harbour Research Institute, Vol. 22, No.2, 1983, 379-42

KAZUAKI Z
"Seawater Corrosion"
Corrosion Control Technology Handbook, Industrial Technology Service Center, 2000, pp 121-134.

LARABEE C P
"New Data Show That Steel Has Low Corrosion Rate During Long Sea Water Exposure"
Materials Protection, 1(12), 95, 1962

LITTLE B J AND WAGNER P A
"The Interrelationship between Marine Biofouling, Cathodic Protection and Microbiologically Influenced Corrosion"
Materials Science Forum, Vols. 192-194, 1995, pp 433-445

MARSH E
"Methods of protection against accelerated low water corrosion"
34th MAFF Conference of River and Coastal Engineers, Keele University,1999

MELCHERS R E
"Effect of Temperature on the Marine Immersion Corrosion of Carbon Steels"
Corrosion, Vol. 58, No 9, September 2002, pp 768-782

MELCHERS R E
"Marine Immersion Corrosion of Mild Steel – An Analyses of World-wide Corrosion Data"
University of Newcastle, Australia, ref. 151.09.1997

MEMET J B, SABOT R, COMPERE C AND DESLOUIS C
"Study of the formation and growth of corrosion products on low carbon steel in natural seawater"
Eurocorr 99

MILLER V G AND OSTROV V
"Replacement of a Deteriorated Steel Sheet Pile Bulkhead"
Ports, 1992

MILLER V G ET AL
"Replacement of a deteriorated steel sheet pile bulkhead"
Proceedings of the Conference Ports'92, Vol. 1, Seattle, Washington, 1992

NUCCI L R AND CONNORS C C
"Inspection and Maintenance of Port Facilities for Improving Service Life"
27th International Navigation Congress, May 1990

PERRY B
"A Concrete Method of Saving Money"
Dredging and Port Construction, September 2001

PRITCHARD A
"Treating Steel Contaminated with MIC"
JPCL, January 2000

REDNER J
"On Microbiologically-influenced Corrosion"
JPCL, November 2001

REPORT OF WORKING GROUP NO 31 OF THE PERMANENT TECHNICAL COMMITTEE 11
"Life Cycle Management of Port Structures General Principles"
International Navigation Association Supplement to Bulletin No 99, 1998

ROBINSON M J AND KILGALLON P J
"Hydrogen Embrittlement of Cathodically Protected High Strength, Low-Alloy Steels Exposed to Sulfate-Reducing Bacteria"
Corrosion, August 1994

SCOTT P B J
"Microbiologically Influenced Corrosion Monitoring: Real World Failures and How to Avoid Them"
Materials Performance, January 2000

SONCK W
"Corrosion to Sheet Steel Pilings: the Paradox of Fresh Water Corrosion"
Port Technology International, No 12, 2001

USACE
"Corrosion and Protection of Steel Piling in Sea Water"
TM 27, 1969

WEST R
"Underwater pile repairs to a jetty in West Africa"
Construction Repair, Vol. 8, No. 4 July/August 1994

G Commercial publications/articles/leaflets

ACOTEC:

- *"Microbiological Corrosion and ALWC,"* March 1999
- *"Accelerated low water corrosion"*, April 2000
- *"The Paradox of Water Level Corrosion of Steel Sheet Piling in a Fresh Water Environment"*, December 2000
- *"Steel Piling Protection in Situ by Humidur: An evaluation"*, February 2001
- *"Corrosion Rate Calculations for ALWC"*, April 2001.

BREAKELL J E, GERRARD A
"Tackling accelerated low water corrosion"
Mott MacDonald /BAC Corrosion Control publication.

CHRISTIE J B
"Steel corrosion and mussel power"
New Civil Engineer 19 October 2000

HODGSON M
"Accelerated low water corrosion – a civil engineer's viewpoint"
Corrosion Management, September/October 1998

HODGSON M
"Limpet dams repair Hamburg quay"
The Structural Engineer, Vol. 76, No 11, June 1998

JOHN MARTIN CONSTRUCTION

- *"A guide to the identification and management of accelerated low water corrosion"*
- *"Rectification, Restoration and Recovery"*
- *"Manual for the Repairing of Sheet Piled and Other Quays Using Limpet Dams"*.

NEW CIVIL ENGINEER:

- *"Steel Bugs Baffle Engineers"*, September 1996
- *"Steel eating bugs spread worldwide"*, March 1998
- *"Warming to limpets"*, March 1998
- *"Why? Is the key question"*, June 1998
- *"Biocorrosion – time for the State to step in"*, October 1998
- *"Biocorrosion fightback awaits alloy trials"*, October 1998

- *"Biocorrosion threat to UK coastal defence"*, October 1998
- *"Port authorities caught out on steel eating bugs"*, October 2000
- *"Ports steel themselves for bugs life horror story"*, October 2000
- *"Beside the sea side"*, September 2003

PARFETT T
"Accelerated low water corrosion (ALWC) Bio Corrosion"
Construction Related Services Engineering Team, Leaflet No. CRS34 Borough of Poole

SIEVEKING P
"Metal-munchers eat piers of the realm"
Corrosion Management, November/December 1998

SONCK W
"Underwater corrosion of steel sheet piling"
Acotec Av, Belgium

Glossary

Active	Freely corroding metal surface; the shiny steel surface seen beneath ALWC patches is indicative of active corrosion.
Active-passive cell	Corrosion cell with one part of a metal surface in the active state acting as an anode and another part of the surface in passive state acting as a cathode.
Aeration cell	See concentration cell; differential aeration.
Aerobe	An organism which can live and grow only in the presence of free oxygen.
Aerobic	Presence of oxygen.
Algae	Primitive one or multi-celled photosynthetic plants; included among algae are kelps and other seaweeds and diatoms.
Anaerobe	An organism for whose life processes a complete or nearly complete absence of oxygen is essential. Sulfate reducing bacteria (SRB) are an example.
Anaerobic	Absence of oxygen.
Anion	A negatively charged ion (such as chloride and sulfate) which migrates towards the anode of a galvanic cell under the influence of a potential difference.
Anode	The electrode at which the anodic reaction predominates and metal loss occurs.
Asset management	A systematic process of maintaining, upgrading and operating physical assets, combining engineering principles with sound business practices and economic theory for the benefit of stakeholders.
Asset register database	A detailed account of the physical extent and properties of a maritime structure established from inspections and used for risk analysis.
Bacteria	Any of numerous unicellular micro-organisms that multiply by fission. Several types of bacteria are known to create corrosive conditions (eg sulfate reducing bacteria, sulfur oxidising bacteria and so on).
Bimetallic corrosion	Corrosion of two metals in electrical contact in an electrolytic solution in which one metal (the cathode) stimulates attack on the other (the anode) and may itself corrode more slowly (be protected) than when it is not in such contact.
Biocide	Chemical substances that destroy bacteriological life. An example of a biocide is hypochlorite generated by electrolysis of seawater (ie chlorination).
Biofilm	Micro-organisms attached to a substrate within a film of extra-cellular polymeric substances released from bacteria.

Calcareous deposit	An alkaline surface deposit consisting largely of calcium carbonate and magnesium hydroxide. Calcareous deposits are formed during cathodic protection and their build up over time is responsible for reducing the protection current required.
Cathode	Electrode at which the cathodic reaction predominates.
Cathodic protection	Electrochemical protection by decreasing the corrosion potential to a level at which the corrosion rate of the metal is significantly reduced. Applied by means of galvanic anodes or an impressed current system.
Cation	A positively charged ion (eg ferrous ion) which migrates towards the cathode of a galvanic cell under the influence of a potential difference.
Chart Datum	The level below which soundings are given on an Admiralty chart and above which tidal levels are measured. In the UK, *Chart Datum* is normally approximately Lowest Astronomical Tide. It is a local datum and does not have a consistent relationship to UK Ordnance Datum Newlyn.
Concentration cell	Corrosion cell in which the potential difference arises from a difference in concentration of the electroactive agents near its electrodes. A common electroactive agent in marine corrosion is dissolved oxygen (see differential aeration).
Condition appraisal	Part of the asset management cycle which follows a continual process of inspection, assessment and, where necessary, remedial treatment, leading to a continual awareness and improvement in infrastructure condition.
Corrosion	The chemical or electrochemical reaction of a metal with its environment, resulting in its progressive degradation or destruction.
Corrosion cell	Short circuited galvanic cell in a corrosion system, the corroding metal forming one of its electrodes. *Macrocells* refer to corrosion cells where the anodes and cathodes are well separated ranging from millimetres (eg as in deposit corrosion) up to several kilometres (eg due to differential aeration on buried pipelines). *Microcells* refer to corrosion where the distance between anodes and cathodes are of microscopic dimensions (eg as in uniform corrosion).
Corrosion product	Substance formed as a result of corrosion.
Corrosion rate	Corrosion effect on a metal per time. Often expressed as an increase in corrosion depth (ie loss of metal thickness) per time as mm/side/year. Corrosion rates may not be linear with time and should be accompanied by information on the type, time dependency and location of the corrosion effect.
Crevice corrosion	Localised corrosion associated with and taking place around a narrow aperture or clearance formed

between the metal surface and another surface (metallic or non-metallic).

Corrosion current Anodic partial current due to metal oxidation.

Current density The current per unit area of the electrode.

Corrosion inhibitor A chemical substance that when added to the environment, generally in small concentrations, decreases the corrosion rate without significantly changing the concentration of any corrosive agent.

Culture media Nutrient mix used for the growth of bacteria (eg API RP38 media).

Datum line A calibrated line used to observe level or depth below a fixed datum.

Design life The period for which a structure or component is designed to be in-service and can economically be maintained rather than replaced. Designers should specify the maintenance regime.

Deposit (or under-deposit) corrosion A form of crevice corrosion associated with and taking place under, or immediately around, a deposit of corrosion product or other substance usually as a result of a differential aeration mechanism.

Differential aeration Corrosion cell in which the potential difference arises from a difference in the concentration of oxygen near its electrodes.

Depreciation rate The rate at which an asset's value is written down for accounting purposes usually referred to as an annual percentage.

Discount rate Rate (annual percentage) at which an asset is depreciated ie rate at which it loses its value.

Doubler plate A plate welded over a sheet pile or tubular pile to either repair an existing hole or in anticipation of future corrosion.

Electrochemical corrosion Corrosion involving at least one anodic reaction and one cathodic reaction.

Electrochemical protection Corrosion protection achieved by electrical control of the corrosion potential eg cathodic protection.

Electrode An electronic conductor in contact with an electrolyte.

Electrode potential Voltage measured in the external circuit between an electrode and a reference electrode in contact with the same electrolyte. The type of reference electrode should always be stated when quoting electrode potentials.

Active potential: a potential towards the negative end of the scale of electrode potentials. The potential of an electrode which is made more cathodic is said to become "more active" ie more negative.

Negative potential; a potential more negative than the potential of the standard hydrogen electrode (ie active).

Noble potential; a potential towards the positive end of a scale of electrode potentials. The potential of an electrode which is made anodic is said to become

	"more noble" ie more positive. *Positive potential*; a potential more positive than the potential of the standard hydrogen electrode (ie noble).
Electrolyte solution	A medium, such as seawater, in which electric current is transported by ions.
Electro-osmosis	The passage of water through a semi-permeable membrane (eg organic coating) under the influence of a potential difference.
Erosion	Destruction of a metal or material by the abrasive action of fluids (eg due to tidal currents) possibly accelerated by the presence of solid matter in suspension (eg sand). When erosion is also combined with a corrosion reaction the term erosion/corrosion is used.
Failure mode and effects analysis	Procedure by which each potential failure mode of a system is analysed to determine the effects on this system and to subsequently classify each potential failure mode according to its severity.
Free corrosion potential	The potential of a metal surface in a given electrolyte solution in the absence of a net (external) electrical current flowing to or from the metal surface.
Galvanic anode	Metallic component used as an anode for galvanic cathodic protection.
Galvanic cell	Combination of different electrodes connected in series with an electrolyte.
Galvanic corrosion	Corrosion due to the action of a corrosion cell. The potential difference is obtained by the electrodes being of different metals, inhomogeneities in a single metal or differences in the contacting solution, such as temperature, oxygen concentration, etc. Note that the term has often in the past been restricted to the action of bimetallic corrosion.
Galvanic series	List of metals ordered according to their free corrosion potentials under specified conditions.
General (or uniform) corrosion	Corrosion or metal loss proceeding at almost the same rate over the whole surface of the metal exposed to the corrosive environment. However, the maximum depth of penetration may still be two to three times the average depth. ALWC is not a form of general/uniform corrosion.
"Green Rust"	Green rusts are obtained when anions such as chloride or sulfate are in excess in a solution after the formation of ferrous hydroxide during corrosion of steel in seawater (see Appendix A1). Typically chloride containing green rust is referred to as "Green Rust 1" and that containing sulfate as "Green Rust 2".
Hazard	An event, process or mechanism that may affect the performance of a structure or present risk to personnel and compromise performance objectives from being met.

Heat affected zone
Region adjacent to weld where the steel has been heated to a high enough temperature to cause microstructural changes but not high enough to melt the steel. The corrosion characteristics of steel in the heat affected zone can differ from parent metal and weld bead and may result in preferential weld corrosion depending upon the steel composition and the welding consumables and conditions.

High yield steel
Steel grade S335GPor higher in accordance with EN 10248.

Holiday
A defect in a coating at which metal is exposed; this may be as small as a pinhole.

Impressed current cathodic protection
Electrochemical protection in which the protection is supplied by an external source eg mains electricity supply via a transformer rectifier (T/R).

Inpan
The flange (pan) of the recessed pile when viewed from the front of a steel sheet pile wall (see Figure 2.4).

Ion
An electrically charged particle which migrates in an electric field.

Inspection
A planned and systematic viewing of an asset, using appropriate destructive or non-destructive testing and photography, to ascertain and record its condition or other characteristics. Used to provide an asset register of condition and hence an estimate of costs for future years. It allows the condition of structures to be compared with the condition of other assets and priorities set. Areas are identified which require more detailed assessment for both operational and safety reasons.

"Instant OFF" potential
The protective potential determined without the IR drop errors when a cathodic protection current is flowing. It is obtained by measurement of the potential momentarily after the applied current has been discontinued to remove the internal resistance seen by the current.

Key performance indicators (KPI's)
KPI's are pre-agreed, quantifiable measurements that permit the objective assessment of assets, methods, contractors and designers performance to reflect the critical performance factors of an asset. They will differ depending on the asset, but may include visual appearance, structural safety, planned remaining service life and others.

Limpet dam
A box-like, temporary structure usually in steel, open to the atmosphere at the top and solid at the bottom used to gain access for dry inspection or repair to maritime structures at any state of the tide. It is likely to be specially designed for a structure and when configured for walls it is solid on three sides and open to the structure on one side. When configured for tubular piles it is usually vertically hinged to allow it to fit concentrically around the tube.

Localised corrosion	Corrosion preferentially concentrated on discrete sites of the metal surface exposed to the corrosive environment. ALWC is an example of localised corrosion.
Low alloy steel	Although definitions vary between different countries, a low alloy steel can be regarded as an alloy steel containing between 1per cent and less than 5 per cent of elements deliberately added for the purpose of modifying properties.
Lowest astronomical tide	The lowest tidal level which can be predicted to occur under average meteorological conditions and under any combination of astronomical conditions. This level will not be reached every year and varies from site to site. In the UK, *Chart Datum* (CD) is approximately LAT.
Maintenance	Work undertaken to keep or restore a structure, its services and site to defined standards.
Marine corrosion	Corrosion with sea water as the main agent of the corrosive environment.
Marine biofouling	Marine organisms such as algae, crustacea, mollusca, etc which form on steel surface.
Mean high water neaps	The average over a year of the heights of two successive high waters at neap tides.
Mean high water springs	The average, over a year of the heights of two successive high waters at spring tides
Mean low water neaps	The average over a year of the heights of two successive low waters at neap tides.
Mean low water springs	The average, over a year of the heights of two successive low waters at spring tides. A more complete definition is provided in the *Mariner's Handbook* published by the Admiralty.
Microbiologically induced corrosion (MIC); also sometimes referred to as "microbiologically influenced corrosion" or "microbially assisted corrosion"	Corrosion caused or accelerated by biological organisms. Refers to the way a wide range of bacteria and micro-organisms stimulate corrosion by direct or indirect activity. It is closely associated with the colonisation of metallic surfaces and the subsequent development of biofilms. ALWC is a form of MIC.
Mild steel	An alloy of iron and carbon with carbon content from 0.1-0.25 per cent. Steel grade S270GP in accordance with EN 10248 for steel sheet piles is the modern equivalent of mild steel.
Neap tides	Tides of relatively small range occurring near the times of the moon's quarters.
Neglect	Inadequacy of condition monitoring or maintenance.
Osmosis	The passage of moisture through a semi-permeable membrane (eg organic coating, marine biofouling layer, etc) from a solution of less concentration to one of greater concentration. This can typically result in blister formation.
Outpan	The flange (pan) of the exposed pile when viewed from the front of a steel sheet pile wall. (see Figure 2.4).

Oxidation	A process in which a reactant loses one or more electrons (see Appendix A1).
pH	A measure of acidity or alkalinity of a solution on a logarithmic scale; a pH <7 is acid, pH=7 is neutral and pH >7 is alkaline.
Pitting corrosion	Localised corrosion resulting in pits in the metal surface ie the formation of cavities extending from the surface into the metal.
Polarisation	Change of electrode potential.
Preferential weld Corrosion	Corrosion associated with the presence of a welded joint and taking place in the weld or its vicinity.
Protective layer	Layer of a substance on a metal surface that decreases the corrosion rate. Such layers may be applied (eg organic coatings) or arise spontaneously (eg corrosion product).
Reduction	An electrochemical reaction in which a reactant accepts one or more electrons.
Reference electrode	An electrode having a stable and reproducible potential that is used as a reference in the measurement of electrode potentials eg standard hydrogen electrode, silver/silver chloride and copper/copper sulfate reference electrodes.
Residual life	The forecast actual (rather than original design) life remaining for an asset, when it will still meet specified design parameters, following reassessment of the condition of the asset and considering its apparent rate of deterioration. The life will always be conditional on a proposed maintenance regime. The residual life of an asset is determined in an engineering assessment based on data gained during an inspection, ie the actual condition of the asset, rather than the difference between (theoretical) design life and the time passed since construction.
Resistivity	Intrinsic property of a conductor which gives the resistance to current flow in terms of its dimensions (length and cross-sectional area).
Risk	The combination of the probability and consequences of a hazard occurring.
Risk assessment	A systematic examination of identifying potential hazards, their probability and the consequence of them occurring. This may apply to an asset, action or method of work. Most notably required by Safety at Work legislation. A full procedure would also list preventative action taken and propose further improvements. In relation to ALWC, it would assess its likelihood, severity and consequences both functional and financial.
Risk mitigation	Measures taken either by reducing/removing a hazard or to minimising its likelihood so that the residual risk is optimal, eg by monitoring, increased maintenance and remedial action.

Risk register	A list of the risks arising from relevant hazards and the costs and benefits of mitigating them.
Rust	Visible corrosion products on ferrous metal surfaces consisting mainly of hydrated iron oxides. Normal rust constituents are typically seen on the steel surfaces in the atmospheric and tidal zones comprising an orange/brown outer layer (ie formed due to unrestricted oxygen access) and a black inner layer (ie formed under limited oxygen conditions).
Scale	Solid layer of corrosion products.
Serviceability limit state	The state of deterioration of a structure such that its use is affected, its durability is impaired or its maintenance requirements are substantially increased. The Serviceability Limit State (SLS) also defines the level of parameters such as stress, deflection or movement when the asset ceases to function properly or safely. The functional use of a structure is disrupted, but collapse is not expected. The Ultimate Limit State (ULS) defines collapse conditions. Both may incorporate a Factor of Safety. Often the first sign of ALWC in earth retaining structures is a surface depression caused by loss of fill through holes in the wall; this indicates that the SLS has been passed and the ULS could be imminent.
Skeletal structure	A structure formed of individual structural members, ie not a solid structure
Spring tides	Tides of relatively large range occurring near the times of new or full Moon.
Stray current corrosion	Corrosion caused by current flowing through paths other than the intended circuits. It may occur for example due to current leakage from DC cables or from induction caused from dockside cranes.
Tubercle	A small rounded prominence on a metal surface due to the local build up of deposit resulting from the enclosure of micro-organisms by iron oxide, iron sulfide or cellular material.
U (Larssen) pile	Steel sheet pile formed by creating interlocks at the edges of a "U" profile. When U piles are joined together the interlocks are all located on the central axis of the resulting wall. (see Figure 2.4).
Ultimate limit state	State of collapse, instability or forms of failure that may endanger property or people or cause major economic loss.
Water-line corrosion	Corrosion along, and as a consequence of the presence of, a gas/liquid boundary. An example of differential aeration.
Z (Frodingham) pile	Steel sheet pile formed by creating interlocks at the edges of a "Z" profile. When Z piles are joined together the interlocks are all located alternately on the front and back faces of the resultant wall. (see Figure 2.4).

For further definitions and information, the reader should refer to:

- ISO Standard 8044 "*Corrosion of metals and alloys – basic terms and definitions*"
- admiralty tide tables
- technical dictionaries including "*Collins Civil Engineering Dictionary*".